Candida albicans

METHODS IN MOLECULAR BIOLOGY™

John M. Walker, SERIES EDITOR

502. **Bacteriophages:** *Methods and Protocols, Volume 2: Molecular and Applied Aspects*, edited by *Martha R. J. Clokie and Andrew M. Kropinski* 2009

501. **Bacteriophages:** *Methods and Protocols, Volume 1: Isolation*, Characterization, and Interactions, edited by *Martha R. J. Clokie and Andrew M. Kropinski* 2009

499. **Candida albicans:** *Methods and Protocols*, edited by *Ronald L. Cihlar and Richard A. Calderone, 2009*

496. **DNA and RNA Profiling in Human Blood:** *Methods and Protocols*, edited by *Peter Bugert, 2009*

493. **Auditory and Vestibular Research:** *Methods and Protocols*, edited by *Bernd Sokolowski, 2009*

490. **Protein Structures, Stability, and Interactions**, edited by *John W. Schriver, 2009*

489. **Dynamic Brain Imaging:** *Methods and Protocols*, edited by *Fahmeed Hyder, 2009*

485. **HIV Protocols:** *Methods and Protocols*, edited by *Vinayaka R. Prasad and Ganjam V. Kalpana, 2009*

484. **Functional Proteomics:** *Methods and Protocols*, edited by *Julie D. Thompson, Christine Schaeffer-Reiss, and Marius Ueffing, 2008*

483. **Recombinant Proteins From Plants:** *Methods and Protocols*, edited by *Loïc Faye and Veronique Gomord, 2008*

482. **Stem Cells in Regenerative Medicine:** *Methods and Protocols*, edited by *Julie Audet and William L. Stanford, 2008*

481. **Hepatocyte Transplantation:** *Methods and Protocols*, edited by *Anil Dhawan and Robin D. Hughes, 2008*

480. **Macromolecular Drug Delivery:** *Methods and Protocols*, edited by *Mattias Belting, 2008*

479. **Plant Signal Transduction:** *Methods and Protocols*, edited by *Thomas Pfannschmidt, 2008*

478. **Transgenic Wheat, Barley and Oats:** *Production and Characterization Protocols*, edited by *Huw D. Jones and Peter R. Shewry, 2008*

477. **Advanced Protocols in Oxidative Stress I**, edited by *Donald Armstrong, 2008*

476. **Redox-Mediated Signal Transduction:** *Methods and Protocols*, edited by *John T. Hancock, 2008*

475. **Cell Fusion:** *Overviews and Methods*, edited by *Elizabeth H. Chen, 2008*

474. **Nanostructure Design:** *Methods and Protocols*, edited by *Ehud Gazit and Ruth Nussinov, 2008*

473. **Clinical Epidemiology:** *Practice and Methods*, edited by *Patrick Parfrey and Brendon Barrett, 2008*

472. **Cancer Epidemiology, Volume 2:** *Modifiable Factors*, edited by *Mukesh Verma, 2008*

471. **Cancer Epidemiology, Volume 1:** *Host Susceptibility Factors*, edited by *Mukesh Verma, 2008*

470. **Host-Pathogen Interactions:** *Methods and Protocols*, edited by *Steffen Rupp and Kai Sohn, 2008*

469. **Wnt Signaling, Volume 2:** *Pathway Models*, edited by *Elizabeth Vincan, 2008*

468. **Wnt Signaling, Volume 1:** *Pathway Methods and Mammalian Models*, edited by *Elizabeth Vincan, 2008*

467. **Angiogenesis Protocols:** *Second Edition*, edited by *Stewart Martin and Cliff Murray, 2008*

466. **Kidney Research:** *Experimental Protocols*, edited by *Tim D. Hewitson and Gavin J. Becker, 2008*

465. **Mycobacteria, Second Edition**, edited by *Tanya Parish and Amanda Claire Brown, 2008*

464. **The Nucleus, Volume 2:** *Physical Properties and Imaging Methods*, edited by *Ronald Hancock, 2008*

463. **The Nucleus, Volume 1:** *Nuclei and Subnuclear Components*, edited by *Ronald Hancock, 2008*

462. **Lipid Signaling Protocols**, edited by *Banafshe Larijani, Rudiger Woscholski, and Colin A. Rosser, 2008*

461. **Molecular Embryology:** *Methods and Protocols, Second Edition*, edited by *Paul Sharpe and Ivor Mason, 2008*

460. **Essential Concepts in Toxicogenomics**, edited by *Donna L. Mendrick and William B. Mattes, 2008*

459. **Prion Protein Protocols**, edited by *Andrew F. Hill, 2008*

458. **Artificial Neural Networks:** *Methods and Applications*, edited by *David S. Livingstone, 2008*

457. **Membrane Trafficking**, edited by *Ales Vancura, 2008*

456. **Adipose Tissue Protocols, Second Edition**, edited by *Kaiping Yang, 2008*

455. **Osteoporosis**, edited by *Jennifer J. Westendorf, 2008*

454. **SARS- and Other Coronaviruses:** *Laboratory Protocols*, edited by *Dave Cavanagh, 2008*

453. **Bioinformatics, Volume 2:** *Structure, Function, and Applications*, edited by *Jonathan M. Keith, 2008*

452. **Bioinformatics, Volume 1:** *Data, Sequence Analysis, and Evolution*, edited by *Jonathan M. Keith, 2008*

451. **Plant Virology Protocols:** *From Viral Sequence to Protein Function*, edited by *Gary Foster, Elisabeth Johansen, Yiguo Hong, and Peter Nagy, 2008*

450. **Germline Stem Cells**, edited by *Steven X. Hou and Shree Ram Singh, 2008*

449. **Mesenchymal Stem Cells:** *Methods and Protocols*, edited by *Darwin J. Prockop, Douglas G. Phinney, and Bruce A. Brunnell, 2008*

448. **Pharmacogenomics in Drug Discovery and Development**, edited by *Qing Yan, 2008*

447. **Alcohol:** *Methods and Protocols*, edited by *Laura E. Nagy, 2008*

446. **Post-translational Modifications of Proteins:** *Tools for Functional Proteomics, Second Edition*, edited by *Christoph Kannicht, 2008*

445. **Autophagosome and Phagosome**, edited by *Vojo Deretic, 2008*

444. **Prenatal Diagnosis**, edited by *Sinhue Hahn and Laird G. Jackson, 2008*

443. **Molecular Modeling of Proteins**, edited by *Andreas Kukol, 2008*

METHODS IN MOLECULAR BIOLOGY™

Candida albicans

Methods and Protocols

Edited by

Ronald L. Cihlar
and
Richard A. Calderone

*Georgetown University Medical Center, School of Medicine,
Department of Microbiology & Immunology, Washington, DC, USA*

Humana Press

Editors
Ronald L. Cihlar
Georgetown University Medical Center
School of Medicine
Department of Microbiology &
  Immunology
Washington DC, USA
cihlarr@georgetown.edu

Richard A. Calderone
Georgetown University Medical Center
School of Medicine
Department of Microbiology &
  Immunology
Washington DC, USA
calderor@georgetown.edu

Series Editor
John M. Walker
University of Hertfordshire
Hatfield, Herts.
UK

ISSN 1064-3745   e-ISSN 1940-6029
ISBN 978-1-58829-760-0   e-ISBN 978-1-60327-151-6
DOI 10.1007/978-1-60327-151-6

Library of Congress Control Number: 2008942152

© Humana Press, a part of Springer Science+Business Media, LLC 2009
All rights reserved. This work may not be translated or copied in whole or in part without the written permission of the publisher (Humana Press, c/o Springer Science + Business Media, LLC, 233 Spring Street, New York, NY 10013, USA), except for brief excerpts in connection with reviews or scholarly analysis. Use in connection with any form of information storage and retrieval, electronic adaptation, computer software, or by similar or dissimilar methodology now known or hereafter developed is forbidden.
The use in this publication of trade names, trademarks, service marks, and similar terms, even if they are not identified as such, is not to be taken as an expression of opinion as to whether or not they are subject to proprietary rights.

Printed on acid-free paper

springer.com

Preface

This book is designed to serve researchers as a source book for methodologies related to the study of medically important fungi and *Candida spp.*, in particular. We have followed the organization of previous volumes in this series in regard to the presentation of each chapter. The past decade has witnessed numerous advances in the study of human pathogenic fungi in areas of biochemistry, molecular biology, taxonomy, and physiology. In addition, the availability of genome sequences of pathogens such as *Candida albicans, Cryptococcus neoformans, Aspergillus fumigatus*, and other model fungi has resulted in new, exciting insights into the pathogenesis of fungal diseases Thus, chapter contributions in this volume have been selected to provide the reader with a variety of approaches that cross discipline lines. For example, because of the critical importance of molecular methods, we have included chapters on reporter gene assays, transformation, gene expression *in vivo*, and methods for large-scale gene disruption. Chapters concerning preparation of samples for proteomic investigations as well as tandem affinity purification, which allow for the identification of interacting proteins, have also been included. The latter chapter highlights the beginning of our understanding of how genes (as words in a sentence) can be organized into a higher level of complexity so that words (genes and proteins) can be arranged into sentences (interacting genes/proteins). Methods for the study of immune response to fungal infections are highlighted in chapters on the evaluation of candidate vaccines, SIgA in protection, the interaction of fungi with dendritic cells, and phagocyte assays with fungi. Likewise, strain identification is vital in studies of pathogenesis and in clinical settings. This topic is discussed in chapters that provide these determinations by DNA fingerprinting or sensitivity to killer toxins. Finally, disease models of candidiaisis are described, and these discuss animal models as well as *in vitro* models (biofilm and tissue culture) that evaluate virulence. The text does not attempt to be inclusive for every current method, but rather the protocols most used are discussed. However, chapters reference alternative procedural approaches, and it is anticipated that the text will well serve the investigators as a source of methods in the field of medical and molecular mycology.

Ronald L. Cihlar, PhD
Richard A. Calderone, PhD

Contents

Preface v
Contributors ix

PART I IMMUNOGICAL METHODS

1 Isolation of Dendritic Cells from Human Blood for *In Vitro* Interaction Studies with Fungal Antigens 3
   T. Sreevalsan

2 Detection and Quantitation of Antifungal SIgA Antibodies in Body Fluids 9
   Michael F. Cole

3 Phagocytosis and Killing Assays for *Candida* Species 17
   Chen Du and Richard A. Calderone

4 Immunization Protocols for Use in Animal Models of Candidiasis 27
   Esther Segal and Hana Sandovsky-Losica

PART II VIRULENCE AND BIOFILMS

5 Penetration of Antifungal Agents Through *Candida* Biofilms 37
   L. Julia Douglas

6 *Candida* Biofilm Analysis in the Artificial Throat Using FISH 45
   Bastiaan P. Krom, Kevin Buijssen, Henk J. Busscher,
   and Henny C. van der, Mei

7 Conditions for Optimal *Candida* Biofilm Development in Microtiter Plates ... 55
   Bastiaan P. Krom, Jesse B. Cohen, Gail McElhaney-Feser, Henk J. Busscher,
   Henny C. van der Mei, and Ronald L. Cihlar

PART III VIRULENCE MEASUREMENTS: IN VITRO, EX VIVO, AND IN VIVO

8 Animal Models of Candidiasis 65
   Cornelius J. Clancy, Shaoji Cheng and Minh Hong Nguyen

9 *Candida albicans* Gene Expression in an *In Vivo* Infection Model 77
   Michael D. Kruppa

10 *In Vitro* and *Ex Vivo* Assays of Virulence in *Candida albicans* 85
   Richard A. Calderone

PART IV STRAIN TYPING AND IDENTIFICATION

11 Biotyping of *Candida albicans* and Other Fungi by Yeast Killer Toxins Sensitivity 97
   Luciano Polonelli and Stefania Conti

12 DNA Fingerprinting *Candida* Species 117
   Claude Pujol and David R. Soll

PART V GENOMICS AND PROTEOMICS

13 The Application of Tandem-Affinity Purification to *Candida albicans* 133
    Chris Blackwell and Jeremy D. Brown

14 Preparation of Samples for Proteomic Analysis of the *Candida albicans*
    Cell Wall 149
    Neeraj Chauhan

15 Reporter Gene Assays in *Candida albicans* 157
    Joy Sturtevant

16 Genetic Transformation of *Candida albicans* 169
    Ana M. Ramon and William A. Fonzi

17 Large-Scale Gene Disruption Using the *UAU1* Cassette 175
    Clarissa J. Nobile and Aaron P. Mitchell

PART VI APPENDIX

18 Standard Growth Media and Common Techniques for Use
    with *Candida albicans* 197
    Neeraj Chauhan and Michael D. Kruppa

Index 203

Contributors

CHRIS BLACKWELL • *Institute for Cell and Molecular Biosciences, The Medical School, Newcastle University, Newcastle upon Tyne, United Kingdom*
JEREMY D. BROWN • *Institute for Cell and Molecular Biosciences, The Medical School, Newcastle University, Newcastle upon Tyne, United Kingdom*
KEVIN BUIJSSEN • *Department of Biomedical Engineering and Departments of Otorhinolaryngology, University Medical Center Groningen and the University of Groningen, Groningen, The Netherlands*
HENK J. BUSSCHER • *Department of Biomedical Engineering, University Medical Center Groningen and the University of Groningen, Groningen, The Netherlands*
RICHARD A. CALDERONE • *Department of Microbiology and Immunology, Georgetown University Medical Center, Washington, DC, USA*
NEERAJ CHAUHAN • *Department of Microbiology and Immunology, Georgetown University Medical Center, Washington, DC, USA*
SHAOJI CHENG • *University of Florida College of Medicine and North Floirda/South Georgia Veterans Health System, Gainsville FL, USA; Department of Medicine, University of Pittsburgh, Pittsburgh, PA*
RONALD L. CIHLAR • *Department of Microbiology and Immunology, Georgetown University Medical Center, Washington, DC, USA*
CORNELIUS J. CLANCY • *University of Florida College of Medicine and North Florida/South Georgia Veterans Health System, Gainsville, FL, USA; Department of Medicine, University of Pittsburgh, Pittsburgh, PA, USA*
JESSE B. COHEN • *Department of Biology, Georgetown University Medical Center, Washington, DC, USA*
MICHAEL F. COLE • *Department of Microbiology and Immunology, Georgetown University Medical Center, NW, Washington, DC, USA*
STEFANIA CONTI • *Dipartimento di Patologia e Medicina di Laboratorio, Sezione di Microbiologia, Università degli Studi di Parma, Parma, Italy*
L. JULIA DOUGLAS • *Division of Infection and Immunity, Institute of Biomedical and Life Sciences, University of Glasgow, Glasgow, United Kingdom*
CHEN DU • *Department of Microbiology and Immunology, Georgetown University Medical Center, Washington DC, USA*
WILLIAM A. FONZI • *Department of Microbiology and Immunology, Georgetown University Medical Center, Washington DC, USA*
BASTIAAN P. KROM • *Department of Biomedical Engineering, University Medical Center Groningen and the University of Groningen, Groningen, The Netherlands*
MICHAEL D. KRUPPA • *Department of Microbiology, East Tennessee State University, James H. Quillen College of Medicine, Johnson City, TN, USA*

GAIL MCELHANEY-FESER • *Department of Microbiology and Immunology, Georgetown University Medical Center, Washington DC, USA*

AARON P. MITCHELL • *Department of Microbiology Columbia University, New York, NY, USA*

MINH HONG NGUYEN • *University of Florida College of Medicine and North Florida/South Georgia Veterans Health System, Gainsville FL, USA; Department of Medicine, University of Pittsburgh, Pittsburgh, PA, USA*

CLARISSA J. NOBILE • *Department of Microbiology Columbia University, New York, NY 10032, USA; Biological Sciences Program, Department of Biological Sciences, Columbia University, New York, NY, USA*

LUCIANO POLONELLI • *Dipartimento di Patologia e Medicina di Laboratorio, Sezione di Microbiologia, Universitá degli Studi di Parma, Parma, Italy*

CLAUDE PUJOL • *Department of Biological Sciences, The University of Iowa, Iowa City, IA, USA*

ANA M. RAMON • *Department of Microbiology and Immunology, Georgetown University Medical Center, Washington, DC, USA*

HANA SANDOVSKY-LOSICA • *Department of Human Microbiology Sackler School of Medicine, Tel-Aviv University, Tel-Aviv, Israel*

ESTHER SEGAL • *Department of Human Microbiology, Sackler School of Medicine, Tel-Aviv University, Tel-Aviv, Israel*

DAVID R. SOLL • *Department of Biological Sciences, The University of Iowa, Iowa City, IA, USA*

T. SREEVALSAN • *Department of Microbiology and Immunology, Georgetown University Medical Center, Washington DC, USA*

JOY STURTEVANT • *Department of Microbiology, Immunology and Parasitology, Louisiana State University School of Medicine, New Orleans, LA, USA*

HENNY C. VAN DER MEI • *Department of Biomedical Engineering, University Medical Center Groningen and the University of Groningen, Groningen, The Netherlands*

Part I
Immunogical Methods

Chapter 1

Isolation of Dendritic Cells from Human Blood for *In Vitro* Interaction Studies with Fungal Antigens

T. Sreevalsan

Abstract

A method is described to generate dendritic cells (DCs) from human peripheral blood mononuclear cells (PBMCs). The procedure involves two major steps: (1) preparation of monocytes from human PBMCs and (2) *in vitro* differentiation of the monocytes into DCs by growth factors and cytokines. Cells obtained in this fashion are screened for the presence or absence of antigenic markers characteristic of DCs by flow cytometry.

Key words: Dendritic cells, monocytes.

1. Introduction

Dendritic cells (DCs) play a key role in the initiation of the primary immune response in a host following infection by a pathogen *(1, 2)*. These cells are found in the body in two forms: an immature and a mature form. Immature cells are present in every tissue and are efficient in processing naive antigens. Internalization and processing of an antigen by immature DCs lead to their migration to draining lymphoid organs where they undergo maturation. Mature DCs, though inefficient in antigen processing, can interact avidly with resting CD4 and CD8 T-cells resulting in their activation. The maturation of DCs is accompanied by upregulation of CD80, CD86, adhesion molecules, cytokines, chemokines, and their receptors. These cells vary in their ability to transmit regulatory signals. DCs display considerable heterogeneity based on their developmental origin and interactions with various pathogens *(3)*. Recently DCs have been used as a natural adjuvant in vaccination and immunotherapy *(4)*. Many studies using fungal cells have been reported indicating that DCs play an important role in the host defense against infections *(5, 6)*. DCs act *in vivo* early during infection and mediate the uptake and the presentation of antigens to T-cells. These cells appear to connect the process of innate immunity with cell-mediated immunity in fungal pathogenesis *(7)*. Many methods have been described in the literature to generate DCs on a large scale *(8–10)*. In this regard, the technique reported by Danciger et al. *(10)* to prepare human DCs for studying *in vitro* interactions with fungal antigens has proven useful and is described below. The generation of DCs from human PBMCs by this technique involves two major steps: (1) isolation of pure monocytes from PBMCs and (2) differentiation of the monocytes into immature or mature DCs, which is described below.

2. Materials

2.1. Reagents

1. Heparin (Sigma),
2. Ficoll-Paque Plus and percoll (Amersham Biosciences),
3. Complete medium (CM): Iscove's modified Dulbeccos's medium (IMDM) (Invitrogen), 25 mM HEPES without phenol red (Life Technologies),
4. Fetal bovine serum (FBS) (Hyclone),
5. Dulbecco's phosphate-buffered saline, 1 mM EDTA, pH 7.2 (DPBS),
6. Cytospin centrifuge (Shandon Elliot),
7. Protocol Hema 3 (Fisher).
8. The following reagents can be obtained from Becton Dickinson: anti-CD1a, anti-DCISGN, anti-CD14 and appropriate isotype sera, paraformaldehyde, and NaN_3 and
9. FACS buffer: 2% FBS, 1% NaN_3 in PBS lacking Ca^{++} or Mg^{++}.

3. Methods

3.1. Isolation of Monocytes

1. From healthy, fasting human volunteers, 360 mL blood is drawn into a container to which heparin (3.8 units/mL) has been added.
2. Aliquots of blood (30 mL) are removed to 50 mL polypropylene tubes (*see* **Note 1**).

3. Then, 6 mL of DPBS is added to each tube.
4. The blood is then underlayered with 10 mL of Ficoll-Paque Plus.
5. Centrifuge at $400 \times g$ for 20 min at room temperature (RT). Leave the centrifuge break-off during this step.
6. A dense white band seen just above the red blood cells is removed carefully with a plastic 5 mL pipette. The banded cells are pooled in 15 mL aliquots. Fractions contain the PBMCs.
7. PBMC-containing aliquots (15 mL) are diluted with 35 mL of DPBS in siliconized centrifuge tubes (see **Notes 2 and 3**). The diluted cell suspension is centrifuged at $150 \times g$ for 10 min at RT. Platelets remain in the supernatant after this centrifugation. Each pellet is resuspended in 10 mL of DPBS and pooled.
8. The above step is repeated again to remove any remaining platelets.
9. Pellets (from four original aliquots) are pooled and suspended in 40 mL CM. Cell concentration is determined by counting in a hemacytometer.
10. The cells are diluted with CM (containing no phenol red) to a concentration of $1-2 \times 10^6$ in a sterile siliconized 500 mL flask. Aliquots of 25 mL are placed in 50 mL polypropylene centrifuge tubes.
11. The suspended cells are underlayered slowly and evenly with 25 mL of CM containing 46% iso-osmotic percoll (see **Note 4**). A 60 mL syringe and a long needle are used for the underlayering task.
12. The resulting discontinuous gradient is centrifuged at $550 \times g$ for 30 min at RT with the break off. After centrifugation, lymphocytes are found in the pellet, while the monocytes are found in the white band formed at the interface between the lower and upper portions of the gradient.
13. Monocytes are collected from the white band using a pipette. Bands from several such gradients are pooled. Aliquots of 15 mL from the pooled bands are diluted with 35 mL of ice-cold DPBS in 50 mL siliconized glass tubes that are on ice.
14. The cell suspension is centrifuged at $400 \times g$ for 10 min at 4°C.
15. Each pellet is resuspended in 20.5 mL of DPBS without EDTA.
16. An appropriately sized aliquot is used to determine cell density using a hemacytometer.

3.2. Monocyte Differentiation into Dendritic Cells

Differentiation of monocytes into immature DCs can be accomplished by incubating them with the appropriate stimulating factor under study by the investigator. Likewise, immature or mature DCs are chosen to fit the investigation. For example, a combination of granulocyte-macrophage colony–stimulating factor (GM-CSF) and interleukin-4 (IL-4) produce immature DCs *(11)*. Such immature DCs also express CD1a on their surface. After stimulation with other agents like lipopolysaccharide (LPS) or a combination of proinflammatory cytokines (IL-1beta, TNF-alpha, PGE2, and monocyte-conditioned medium (MCM)) (*see* **Note 5**), the immature DCs show increased expression of CD86, CD80, and MHC class II molecules, thus representing the mature DCs.

3.2.1. Immature DCs

1. The monocyte cell suspension prepared as described earlier is diluted with CM to a concentration of 6×10^6/mL.
2. Cell suspension of 25 mL is placed in T-150 tissue culture flasks (Corning, NY) and incubated at 37°C for 2 h in order to allow cell adherence.
3. Gentle shaking of the flasks for a few minutes will allow the nonadherent cells to be released to the supernatant medium.
4. The medium is decanted.
5. The cell monolayer is washed with CM.
6. Fresh medium (35 mL) containing IL-4 at 1000 U/mL and GM-CSF (Schering-Plough) at 1000 U/mL are added and incubated at 37°C.
7. After 6 days incubation, the medium in the cultures is replenished with a fresh preparation of growth factors.
8. Two days later, the medium is decanted and nonadherent cells are removed from the cultures and saved.
9. The monolayers are washed gently with PBS to remove loosely adhered cells, leaving behind undifferentiated cells.
10. This fraction is combined with the first batch of nonadherent cells.
11. The cell suspension is centrifuged at $200 \times g$ for 10 min, washed with PBS as before, and resuspended in CM at the desired concentration. This represents the immature DCs.

3.2.2. Mature DCs

1. **Steps 1–6** are identical to that described in **Section 3.2.1**.
2. After 8 days incubation, CM contains LPS or a cocktail of IL-beta, TNF-alpha, PGE2, and MCM.
3. The remainder of the procedure regarding handling the culture is identical to those described for obtaining immature DCs.

3.3. Confirmatory Evidence for DCs Differentiation and Maturation

Differentiation of monocytes into DCs is accompanied by upregulation of antigens CD1a and DCSIGN and a loss of CD14 expression. Thus, the DCs generated from monocytes should be screened for the above antigens using corresponding antibodies by flow cytometry.

1. Detached cells are suspended in 100 µL of FACS buffer.
2. Cells are incubated for 30 min on ice, following the addition of 5 µL conjugated anti-CD1a, anti-DCISGN, or anti-CD14, in the appropriate isotype serum (Becton Dickinson).
3. Samples are washed with FACS buffer.
4. Samples are fixed with 2% paraformaldehyde (Becton Dickinson) and stored at 4°C.
5. Samples are analyzed in a FACS flow cytometer (Becton Dickinson).

3.4. Continuous DC Cell Lines

A few cell lines displaying the phenotypic and functional properties of DCs have been described in the literature *(12–14)*. Upon differentiation, these cells seem to acquire some properties of DCs. Recently, such cell lines have been compared with primary DCs with respect to their ability to undergo both maturation and the accompanying functional changes *(15)*. These cell lines were derived originally from human myeloid leukemia. The cell lines THP-1, KG-1, and MUTZ-3 were studied in this context. The results indicated that MUTZ-3 cells have comparable abilities in functional and transcriptional activity to those of the primary DCs. Therefore, MUTZ-3 cells can act as a model for primary DCs. The ease of cultivation of these cells in the laboratory makes them an ideal system to study the role of DCs in immune regulation.

4. Notes

1. Only sterile glassware or plasticware is used throughout the procedure.
2. Glassware can be siliconized by the following procedure: 50 mL glass centrifuge tubes (Kimax) are heated at 180°C for 4 h to inactive endotoxin, if any, and then cooled; the glassware is then coated with Sigmacote (Sigma).
3. Polypropylene tubes, Blue Max (Falcon) can be substituted for siliconized glassware.
4. Iso-osmotic percoll can be prepared as follows: CM with 46% percoll can be prepared by combining 46 mL of iso-osmotic percoll with 54 mL of CM. Phenol red is included in the CM to give contrast with the layer containing PBMC in medium

lacking phenol red. Iso-osmotic percoll is obtained by combining 9.25 mL of percoll with 0.75 mL of 10X DPBS.

5. MCM can be prepared by incubating the monocyte culture *(3–13)* in CM for 24 h at a cell density of 2 × 10^6 cells/mL at 37°C.

References

1. Stienman, R.M. (1991) The dendritic cell system and its role in immunogenicity. *Ann. Rev. Immunol.* **9**, 271–296.
2. Hart, D.N. (1997) Dendritic cells: Unique leukocyte populations which control the primary immune response. *Blood* **90**, 3245–3287.
3. Shortman, K., and Liu, Y. (2002) Mouse and human dendritic cell subtypes. *Nat. Rev. Immunol.* **2**, 151–161.
4. Fong, L., and Engelman, E.G. (2000) Dendritic cells in cancer immunotherapy. *Ann. Rev. Immunol.* **18**, 245–273.
5. Fe d'Ostiani, C., Del Sero, G., Bacci, A., Montagnoli, C., Spreca, A., Mencacci, A., Ricciardi-Casegnoli, P., and Romani, L. (2000) Dendritic cells discriminate between yeasts and hyphae of the fungus *Candida albicans*: Implications for T helper cell immunity *in vitro* and *in vivo*. *J. Exp. Med.* **191**, 1661–1674.
6. Bacci, A., Montagnoli, C., Perruccio, K., Bozza, S., Gaziano, R., Pitzurra, L., Velardi, A., Fe d'Ostiani, C., Cutler, J.E., and Romani, L. (2002) Dendritc cells pulsed with fungal RNA induce protective immunity to *Candida albicans* in hemapoietic transplantation. *J. Immunol.* **168**, 2904–2913.
7. Romani, L., Montagnoli, C., Bozza, S., Perruccio, K., Spreca, A., Allavena, P., Verbeek, S., Calderone, R.A., Bistoni, F., and Puccetti, P. (2004) The exploitation of distinct recognition receptors in dendritic cells determines the full range of host immune relationships with *Candida albicans*. *Int. Immunol.* **16**, 149–161.
8. Romani, N., Reider, D., Heuer, M., Ebner, S., Kampgen, E., Eibl, B., Neiderwiesser, D., and Schuler, G. (1996) Generation of mature dendritic cells from human blood: An improved method with special regard to clinical applicability. *J. Immunol. Methods* **196**, 137–151.
9. Tuyaerts, S., Noppe, S.M., Corthals, J., Breckpot, K., Heirman, C., Greef, C.D., Riet, I.V., and Thielemans, K. (2002) Generation of large numbers of dendritic cells in a closed system using cell factories. *J. Immunol. Methods* **264**, 135–151.
10. Danciger, J.S., Lutz, M., Hamma, S., Cruz, D., Castrillo, A., Lazaro, J., Phillips, R., Premack, B., and Berliner, J. (2004) Method for large scale isolation, culture and cryopreservation of human monocytes suitable for chemotaxis, cellular adhesion assays, macrophage and dendritic cell differentiation. *J. Immunol. Methods* **288**, 123–134.
11. Bender, A., Sapp, M., Schuler, G., Steinman, R.M., and Bhardwaj, N. (1996) Improved methods for the generation of dendritic cells from nonproliferating progenitors in human blood. *J. Immunol. Methods* **196**, 121–135.
12. Koski, G.K., Schwartz, G.N., Weng, D.E., Czernieki, B.J., Carter, C., Gress, R.E., and Cohen, P.A. (1999) Calcium mobilization in human myeloid cells results in acquisition of individual dendritic cell-like characteristics through discrete signaling pathways. *J. Immunol.* **163**, 82–92.
13. St. Louis, D.C., Woodcock, J.B., Franozo, G., Blair, P.J., Carlson, L.M., Murillo, M., Wells, M.R., Williams, A.J., Smoot, D.S., Kanshal, S., Grimes, J.L., Harlan, D.M., Chute, J.P., June, C.H., Soebelist, U., and Lee, K.P. (1999) Evidence for distinct intracellar signaling pathways in CD34$^+$ progenitor to dendritic cells5 RRRRRR differentiation from a human cell model. *J. Immunol.* **162**, 3237–3248.
14. Ackerman, A.L., and Cresswell, P. (2003) Regulation of MHC class I transport in human dendritic cells and dendritic–like cell line KG-1. *J. Immunol.* **170**, 4178–4188.
15. Larsson, K., Lindstedt, M., and Borrebaek, C.A.K. (2006) Functional and transcriptional profiling of MUTZ-3, a myeloid cell line acting as a model for dendritic cells. *Immunology* **117**, 156–166.

Chapter 2

Detection and Quantitation of Antifungal SIgA Antibodies in Body Fluids

Michael F. Cole

Abstract

The measurement of antibodies in the external secretions that bathe mucosal surfaces is important in understanding the host response to the opportunistic pathogen, *Candida albicans* and its determinants of pathogenesis at these sites. The principal immunoglobulin isotype in mucosal secretions is secretory immunoglobulin A (SIgA). Unlike the circulatory system, mucosal surfaces are open systems in which the concentrations of immune factors are affected by diurnal variation, changes in flow rate, complex formation with mucins, and other variables. Thus, it is necessary to control these factors if meaningful data are to be obtained. This chapter outlines methods for the measurement of anti-*Candida* SIgA antibodies in primary units and shows how to control the factors that influence antibody measurement in external secretions.

Key words: SIgA, SIgA antibodies, *Candida* antigens.

1. Introduction

In studies of the pathogenesis of fungal infections or for the development of vaccines, it is frequently necessary to be able to detect and quantitate antibodies directed against the organism and/or its antigens in various body fluids.

Body fluids can be divided into two types, those in closed systems such as blood, cerebrospinal fluid, peritoneal fluid, etc., and those in open systems, that is, mucosal secretions such as tears, nasal secretions, saliva, milk, genitourinary secretions, etc. While detection of antimicrobial antibodies in blood is generally quite straightforward, the detection and quantitation of antimicrobial antibodies in external secretions is altogether a different proposition. The reasons for this are several and include diurnal variation, the inverse relationship between flow rate and antibody concentration, complexation of antibodies with high-molecular-weight mucins and other factors found in secretions, and proteolytic and gycolytic activity of the resident microbiotas that colonize mucosal surfaces. Detection of antibodies was revolutionized by the invention of the enzyme-linked immunosorbent assay (ELISA) by Engvall and Perlmann *(1)*. This method is very versatile and sensitive, and is widely employed in both research and clinical care settings *(2, 3)*. One limitation of the measurement of antimicrobial antibodies by ELISA is that the data are output in optical density units. While this may be satisfactory for making comparisons within individual laboratories, it makes comparisons with data from other laboratories difficult as optical density is affected by incubation time, source of antibody reagents, nature of the plastic plate, to name but a few. It is the purpose of this chapter to describe the application of ELISA to the detection of anti-*Candida* antibodies in external secretions in which the read-out is in primary units, rather than optical density. The method described combines the measurement of antimicrobial secretory IgA (SIgA) antibodies and total SIgA immunoglobulin on a single 96-well-microtitration plate. The described method is applicable to the detection and measurement of almost all antimicrobial antibodies that are induced in the secretions of experimental animals or humans. The inexperienced reader is advised to familiarize themselves with the basic principles of solid-phase assays before embarking on the assay described in this chapter. A free technical handbook on ELISA and the related technique ELISPOT is available from Pierce Chemical at http://www.piercenet.com/Objects/View.cfm?Type=Page&ID=AF5B61C9-9149-41F4-B63D-DA4C46CD9446

2. Materials

2.1. Collecting Mucosal Secretions

1. One piece 3.0 mL sterile disposable transfer pipette (Fisher Scientific, Pittsburgh, PA, USA).
2. Sterile phosphate-buffered saline (PBS), pH 7.4.
3. 500 mM ethylenediamine tetracetic acid (EDTA) solution. Dissolve 14.61 g of EDTA (Sigma-Aldrich, St. Louis, MO, USA) in 50 mL of deionized-distilled water (ddH_2O) and filter sterilize using a 0.45 μm disposable vacuum filter unit.

2.2. Enzyme-Linked Immunosorbent Assay (ELISA)

1. 96-Well flat bottom microtiter plates (Immulux™ or Immulux™ HB, Dynex Technologies, Chantilly, VA, USA).
2. Coating buffer: 0.05 M carbonate buffer, pH 9.6, containing 0.02% NaN_3. Dissolve 1.6 g of Na_2CO_3 (anhydrous)

(MW = 105.99), 2.9 g of $NaHCO_3$ (MW = 84.01), and 0.2 g of NaN_3 (all from Sigma) in a final volume of 1 L of ddH_2O. Store at 4°C. Discard after 2 weeks.

3. Blocking reagent: 0.1% (1.0 g/L) bovine serum albumin (BSA) fraction V (Sigma) containing 0.02% (0.2 g/L) NaN_3 in PBS, pH 8.0.

4. PBS-Tween: 0.1% (1 mL/L) Tween 20 (polyoxyethylenesorbitan monolaurate, Sigma) containing 0.02% (0.2 g/L) NaN_3 in PBS, pH 8.0.

5. Horseradish peroxidase (HRP)-conjugated antibody diluent: 0.1% (1.0 g/L) BSA in PBS, pH 8.0.

6. Citrate–phosphate buffer: Mix four parts of 0.1 M citric acid (dissolve 1.92 g of anhydrous citric acid (Sigma) in 100 mL of ddH_2O) with six parts of 0.2 M Na_2HPO_4 (dissolve 2.84 g of Na_2HPO_4 (Sigma) in 100 mL ddH_2O). Adjust pH to 4.5, if necessary, with either solution.

7. HRP substrate solution: Dissolve 1 mg of *o*-phenylenediamine (Sigma) in 1.0 mL of citrate–phosphate buffer, pH 4.5 and add 0.012% H_2O_2 (4 μL of 30% H_2O_2 per 10 mL of buffer) (*see* **Note 1**). Prepare substrate solution immediately prior to use. Substrate is light sensitive; use dark bottle or wrap foil around the vessel.

3. Methods

The method described below is designed to quantitate *Candida*-reactive SIgA antibodies in external secretions and output the data in primary units by interpolating optical density into a SIgA standard curve (ng/mL). In this application, *Candida* cells or purified *Candida* antigens form the solid phase, that is, they are immobilized on the plastic surface of the well of one-half of a 96-well-microtiter plate. Here they serve to capture SIgA antibodies reactive with them in the external secretion. On the other half of the plate, an antibody to human secretory component (SC) forms the solid phase and serves to capture a series of dilutions of purified SIgA, thus forming an SIgA standard curve.

3.1. Determination of the Dry Weight of Candida Cells

1. Dry a 25 mm, 0.4 μm Whatman Nuclepore® polycarbonate filter (Fisher Scientific) to constant weight. Place the filter inside a P_2O_5 dessicator, evacuate the dessicator and place it overnight in a 60°C incubator.

2. Weigh individual filters on an analytical balance accurate to 10 μg, and record weight on Petri dish.

3. Set up a 25-mm fritted glass filter base with stopper in 125 mL Erlenmeyer side-arm flask (Millipore XX15 047 00 All-Glass Filter Holder).

4. Place a 0.45-µm nitrocellulose filter (Sigma) on the fritted glass filter base and wet with ddH$_2$O; carefully place the polycarbonate filter on top of the nitrocellulose filter, making sure that there are no air bubbles between the filters.

5. Accurately pipette an aliquot of a *Candida* suspension (~200 µL) onto the central portion of the polycarbonate filter while a vacuum is applied to the side-arm flask.

6. Wash the cells with several hundred microliters of ddH$_2$O.

7. Place the filter back in the same Petri dish, and evacuate the dessicator and warm to 60°C overnight as before.

8. Reweigh the filter, determine weight of cells in the volume applied to the membrane, and calculate the total weight of cells based on the volume of the cell suspension.

3.2. Determination of the Concentration of Antigens

1. The concentration of protein antigens can be determined using commercially available protein assay kits such as the BCA (bicinchoninic acid) Protein Assay or the Coomassie (Bradford) Protein Assay, both obtainable from Pierce Biotechnology, Rockford, IL, USA.

3.3. Collecting Mucosal Secretions

1. Aspirate secretion and dispense into a graduated tube. Measure the volume of the secretion and add sufficient EDTA solution to give a final concentration of 5 mM (*see* **Note 2**).

2. Store the secretion at −70°C until use.

3.4. ELISA to Detect Binding of SIgA Antibodies Against Candida Cells or Antigens (see Note 3)

1. Coat one half of the wells of the microtiter tray (48 wells) with either *Candida* cells or purified antigen by dispensing 100 µL of the stock cell or antigen preparation diluted in coating buffer into each well. The coating concentration routinely used for whole fungal or bacterial cell suspensions is 100 µg dry weight/mL of coating buffer. For purified protein and carbohydrate antigens, the coating concentrations used are usually between 1.0 and 10 µg/mL of coating buffer.
    a. Coat the remaining 48 wells by dispensing 100 µL of a 10 µg/mL dilution of a murine monoclonal antibody to human SC (Hybridoma Labs., Baltimore, MD, USA) in coating buffer.

2. Seal the plate using adhesive plastic film sealer (Fisher).

3. Place microtiter tray on an orbital shaker (Bellco Technology, Vineland, NJ, USA, or similar) and incubate overnight (~16 h) at 4°C.

4. Manually tip out the coating reagent by inverting the plate. Drain residual liquid by tapping the plate, upside down, on a pile of paper towels.

5. Cover all the naked plastic surface in each well with an irrelevant protein. This step is termed 'blocking' (see **Note 4**). To block the wells, wash them three times with blocking reagent filling the entire well (~350 μL). Automatic plate washers are available commercially to perform the wash steps. Alternatively, an 8- or 12-channel multichannel pipette (Eppendorf or equivalent) or a wash bottle may be used. If a wash bottle is used, cut the tip off the spout of the wash bottle such that a gentle stream, rather than a forceful one, is dispensed.

6. Next dispense 300 μL of blocking reagent into each well, seal plate, and incubate for 1 h at room temperature.

7. Manually tip out the blocking reagent by inverting the plate. Drain residual liquid by tapping the plate, upside down, on a pile of paper towels.

8. Dilute secretion samples in PBS-Tween and dispense 100 μL in duplicate or preferably triplicate into the wells coated with the *Candida* cells or antigen (see **Note 5**).

    a. Dispense 100 μL of duplicate or triplicate dilutions of purified SIgA (Cappel-MP Biochemicals, Solon, OH, USA) in PBS-Tween over the range of 10–100 ng/mL. Make dilutions on the day they are to be used.

9. Seal plate and incubate on an orbital shaker at room temperature for 1 h.

10. Manually tip out the content of the wells by inverting the plate. Drain residual liquid by tapping the plate, upside down, on a pile of paper towels.

11. Wash wells three times with PBS-Tween.

12. Make a 1:10,000 dilution of HRP-conjugated affinity purified rabbit IgG antibody to human α-chain (Jackson ImmunoResearch Labs. Inc., West Grove, PA, USA) with HRP diluent (see **Section 2** above) (see **Note 6**). This is often termed as the 'reporter' antibody. Prepare only as much as is required (~10 mL/plate) immediately before use.

13. Dispense 100 μL in each well, seal the plate, and incubate for 1 h on an orbital shaker at room temperature in the dark (cover microtitration plate with aluminum foil).

14. Drain and wash wells as in **steps 9 and 10**. Do not leave PBS-Tween in wells for more than a few minutes because HRP is inactivated by NaN_3.

15. Prepare substrate solution (see **Section 2**) and dispense 100 μL in each well. Seal the plate.

16. Place plate on an orbital shaker and incubate at room temperature.
17. Read optical density at 450 nm in a dedicated microtiter plate spectrophotometer once the colorless substrate turns a yellow-brown color.
    a. Subtract the absorbance of the background control wells (wells that do not contain secretion sample or SIgA standard).
    b. Construct a standard curve by plotting the concentration of SIgA (ng/mL) on the *x*-axis versus optical density (OD_{450}) on the *y*-axis.
18. Take the mean of the duplicate or triplicate optical density readings from the *Candida*-cell- or antigen-coated wells and read them off the SIgA standard curve.

4. Notes

1. The substrate *o*-phenylenediamine is a carcinogen and the powder should be handled with care to prevent aerosolization. Substrate tablets (20 mg) are available from Sigma.
2. The quantitation of antimicrobial SIgA antibodies in external secretions presents a number of issues that are not faced when measuring antimicrobial antibodies in blood. The circulatory humoral immune system is a closed system; that is, immunoglobulin molecules continuously recirculate through the vasculature and interconnected lymphatics, and their concentrations are held at a steady state determined by their rate of degradation (half-life) and rate of new synthesis. In addition, immunoglobulins in blood are not subject to circadian rhythms. In contrast to the closed circulatory system, the mucosa-associated immune system is open, which means that antibody, primarily SIgA, is continuously exported in the local secretions that bathe mucosal surfaces by local exocrine glands. Consequently, the concentrations of SIgA in external secretions are low because the secreted volumes are high. Moreover, the concentration of SIgA in secretions is inversely related to flow rate. Flow rate is influenced by the level of hydration, hormones, drugs, and circadian rhythm. In addition, SIgA forms heterotypic calcium-ion-dependent complexes with mucins and several innate immune components present in secretions. In order to control these variables as much as possible, try to collect secretions at the same time of day. The immediate addition of the chelating agent EDTA upon collection of the secretion inhibits protease activity and

the formation of Ca^{2+}-dependent heterotypic complexes of SIgA with other factors. Whenever possible, measure the flow rate of the secretion or, if this is not possible, normalize the data to the protein concentration of the secretion.

3. The optimal concentrations of each reagent used in the ELISA must be established empirically before executing the assay. Optimal concentrations are determined by performing a 'checkerboard' titration in which the concentrations of two reagents, for example, the capture antibody or antigen and the reporter antibody are varied in a single microtiter tray. Serial dilutions of one reagent are dispensed down columns 1–12 of the plate, and serial dilutions of the second reagent are dispensed across rows A–H. Readers not familiar with performing checkerboard titrations are advised to consult the manual described in the introduction and listed in the references.

   The solid phase, that is, the *Candida* cells, *Candida* antigen, or antibody to human SC, must always be in excess, i.e., they must not be saturated by antibodies in the secretion being tested or by the authentic SIgA that is used to construct the standard curve. This is one such variable that is examined during a checkerboard titration.

   Proteins bind readily to polystyrene from which the microtiter plates are constructed. However, there are cases where the antigen may have to be modified to provide charged groups or the plastic surface treated to permit binding. Microtiter trays having wells specially treated for various applications are available from several commercial sources. Antibodies are useful molecules to 'present' antigens on the plastic surface. Antibodies not only bind well to plastic but also act to hold antigen off the plastic surface, which can be useful to avoid steric hindrance.

   The standard assay volume for ELISA is 100 µL, but volumes between 50 µL and 200 µL can be used depending on scarcity of the sample and the level of antibody in the sample.

4. Blocking the uncoated plastic surface of the well is critical to prevent nonspecific binding of sample or reporter antibodies. As an alternative to BSA used above, which is expensive to purchase, a solution of nonfat dried milk may be used.

   If incubation with antibody is not to be performed the same day as blocking, the plate can be stored at 4°C with the wells filled with blocking reagent.

5. If a large number of samples are to be dispensed into several plates, the incubation period should be increased in order to take account of the difference in incubation time between the first and last samples dispensed.

6. HRP is inactivated by NaN$_3$. Note that the antibody diluent does not contain sodium azide. Although the wash solution does contain azide, the HRP is unaffected by short exposure during washing of the wells.

References

1. Engvall, E., and Permann, P. (1972) Enzyme-linked immunosorbent assay, ELISA. III. Quantitation of specific antibodies by enzyme-linked anti-immunoglobulin in antigen-coated tubes. *J. Immunol.* **109,** 129–135.
2. Cole, M. F., Bryan, S., Evans, M. K., Pearce, C. L., Sheridan, M. J., Sura, P., Wientzen, R., and Bowden, G. H. W. (1999) Humoral immunity to commensal oral bacteria in human infants: Salivary secretory immunoglobulin A antibodies reactive with *Streptococcus mitis* biovar 1, *Streptococcus oralis, Streptoccoccus mutans* and *Enterococcus faecalis* during the first two years of life. *Infect. Immun.* **67,** 1878–1886.
3. Fitzsimmons, S. P., Evans, M. K., Pearce, C. L., Sheridan, M. J., and Cole, M. F. (1994) Immunoglobulin A subclasses in infants' saliva and milk from their mothers. *J. Pediatr.* **124,** 566–573.

Chapter 3

Phagocytosis and Killing Assays for *Candida* Species

Chen Du and Richard A. Calderone

Abstract

Both innate resistance and acquired cell-mediated immunity are involved in an anti-*Candida* response. Essential components of both the arms of the immune defense against infections by *Candida* spp. include phagocytic cells, i.e., polymorphonuclear neutrophils (PMNs) and mononuclear phagocytes. A powerful *in vitro* assay to assess host–pathogen interactions and study pathogenesis is the co-culture of phagocytic cells with a test fungus. The precise contribution of phagocytes to the host defense is usually assessed by determining phagocytosis and killing of *Candida* spp. blastoconidia. Dissection of the roles of various virulence factors in the infection process will involve the use of both *in vitro* and *ex vivo* assays. These assays are very useful as one of the approaches to determine the virulence factors of *Candida spp.*, now that specific gene mutants are relatively easy to construct. *In vitro* studies involving specific cultured immune system cells can permit the analysis of interactions under controlled conditions. These studies provide an opportunity to monitor and compare host cell behavior upon challenge with wild-type or mutant strains of the pathogen.

Key words: Phagocytosis, *ex vivo*, *in vitro*, PMN, monocytes, candicidal activities.

1. Introduction

It is very clear from the literature that *Candida* spp. are important human commensals that may invade mucosal surfaces or cross the mucosal surface resulting in systemic infection. During each of these disease cycles, *Candida* spp. encounter numerous types of immune cells, including neutrophils, monocytes, tissue macrophages, and dendritic cells. These account for the innate immunity against these organisms and also result in the induction of acquired resistance associated with T- and B-cell functions. In this chapter, we focus on the innate cells of the immune system, specifically polymorphonuclear neutrophils (PMNs) and monocytes, and describe assays to assess both their killing capacity and the role of certain fungal genes in *Candida* survival within these cells. But are innate phagocytes equal in their ability to protect the host? The answer to this paradigm, based on both *in vitro* studies with mononuclear and polymorphonuclear phagocytes and clinical observations, is that neutrophils account for more of the innate protective immunity than mononuclear phagocytes *(1–3)*. Given that invasive candidiasis often occurs in neutropenic patients, one might speculate that mononuclear phagocytes are left to protect the host but fail in this task, given the lack of acquired resistance. In the case of *Candida*-PMN studies, both oxidative and nonoxidative mechanisms are critical to killing of the organism *(3)*.

In this regard, the best assay to use in obtaining data about interactions with phagocytes and *Candida* spp. (or any organism) is to use freshly collected PMNs or monocytes from human volunteers. Certainly the volume of cells (and numbers) is useful for doing experiments where, for example, killing of specific strains is to be measured. On the other hand, volunteers are often not available, and variation from experiment to experiment can be a result of using different volunteers or even subtle differences in the same volunteer per collection of his/her blood. For this reason, cell lines of phagocytes are used, more often in regard to mononuclear phagocytes, including a human monocyte cell line (THP-1) and a mouse macrophage cell line (RAW264.7) *(4)*. For studies with neutrophils, only one cell line, HL60, is available, and there are a few reports that describe its preparation and activation *(5–8)*. To reiterate, the use of cell lines decreases the amount of variability in experiments if culture protocols are consistent from experiment to experiment. Second, some experiments require many more cells than can be obtained from volunteers and thus cultured cells can provide the required number of cells.

Below, we describe protocols with PMNs and monocytes and focus on the methods of measuring phagocytosis and killing.

2. Materials

2.1. Isolation of Neutrophils

1. Sodium heparin.
2. 3% Dextran in 0.9% NaCl (dextran–saline solution).
3. Ficoll-Hypaque (can be replaced with Lymphocyte Separation Medium Density = 1.077–0.080 g/mL).
4. Phosphate-buffered saline buffer (PBS): 1 × pH 7.4 without calcium and magnesium (8 g NaCl, 0.2 g KCl, 1.44 g Na_2HPO_4, 0.24 g KH_2PO_4, in 1 L).

5. 0.2% NaCl (washing cells) and 1.6% NaCl, ice-cold, (hypertonic shock to remove RBCs).
6. RPMI-1640 medium, with or without phenol red.
7. 0.4% Trypan blue in PBS. Trypan blue solution.

2.2. Phagocytosis Assay

1. 1.5-mL, 14-mL, and 50-mL conical plastic tubes.
2. 24- and 96-well culture plates (flat bottom).
3. YPD liquid medium: 1% yeast extract, 1% peptone. 2% glucose.
4. Human serum (AB type) (optional).
5. Wright-Giemsa stain.
6. Fluorescein isothiocyanate (FITC): (0.01 mg/mL in 0.5 M carbonate/biocarbonate buffer, pH 9.5).
7. Ethidium bromide (EtBr) (10 mg/mL).

2.3. Killing Assay

1. 2, 3-*bis*(2-methoxy-4-nitro-5-sulfophenyl)-5-[(phenylamino) carbonyl]-2H-tetrazolium hydroxide (XTT), 0.5 mM, pH 7.0.
2. Phenazine methosulfate (PMS), 1% solution.
3. Two commercially available kits: *in vitro* toxicology assay kit XTT based (Sigma); cell proliferation kit II XTT (Roche).

3. Methods

3.1. Phagocytosis Assays

In vitro phagocytosis assays are performed with phagocytic cells/ *Candida* co-cultures on samples removed at specific time points. One method of assessing phagocytosis is by direct microscopic observation *(9)*. An inherent weakness with such an assay, however, is the difficulty of differentiating microscopically ingested yeasts versus those merely bound to the surface of phagocytes. To solve this problem, FITC-labeled *Candida* is used in the co-culture system, followed by fluorescence flow cytometry (FCM) or fluorescence microscopy.

3.1.1. Preparation of Phagocytic Cells (Human)

The protocol described is designed for the isolation of human neutrophils from 50 mL of peripheral blood. With minor modifications, it is suitable for other species and sources.
1. Heparinized blood (20 U/mL) obtained from normal adult volunteers is mixed with an equal volume of dextran/saline solution in a 50-mL tube and incubated in an upright position until a clear interface appears (about 30 min to 1 h).
2. Transfer the leukocyte-rich plasma (upper layer) into a fresh tube and pellet cells from the plasma by centrifuging for 10 min (500 × g).

3. Remove the supernatant and resuspend cells in 30-mL of PBS solution.
4. Using transfer pipettes, gently layer the cell suspension onto a 15 mL Ficoll-Hypaque separation medium in a 50-mL tube and centrifuge for 30 min at 500 × g, with no brake. This can be done at room temperature.
5. Aspirate the upper layer, leaving the neutrophil pellet (which is pelleted along with the remaining RBCs to the bottom of the centrifuge tube).
6. The erythrocytes are lysed by resuspending the cells in 5 mL of ice-cold 0.2% NaCl for 30 s, after which time physiological osmolarity is restored by the addition of 5 mL of ice-cold 1.6% NaCl.
7. Centrifuge for 5 min (600 × g) at 4°C. Discard the supernatant.
8. **Steps 6 and 7** may need to be repeated two to three times to lyse erythrocytes completely (until the red color is no longer visible in the cell pellet).
9. Resuspend cells in 5 mL of RPMI-1640 medium. For determining the cell density and cell viability, mix 15 µL of cell suspensions with 15 µL of trypan blue solution in a 1.5 mL tube, and count cells using a hemacytometer. The cells are ready for immediate use. Otherwise, keep them at 4°C for no more than 6–8 h.

In addition to using freshly isolated phagocytic cells, there are two phagocyte cell lines commonly used in assays.

1. *HL60 cells.* These cells were originally isolated from the peripheral blood of a patient with acute promyelocytic leukemia *(4, 5)*. The immature HL60 cell population is predominantly composed of promyelocytes and a smaller percentage of more mature cells, resembling myelocytes, metamyelocytes, and banded and segmented neutrophils. These cells are neither microbicidal nor tumoricidal, but upon treatment with dimethyl formamide, dimethyl sulfoxide, or retinoic acid, the cells differentiate along the granulocytic pathway characterized by morphological changes, a cessation of proliferation, and the development of a polymorphic nucleus. The cells increase their production of oxidants, and there is also a concomitant increase in their ability to phagocytose pathogens. The HL60 myelomonocytic cell line has been used as an *in vitro* model for studying neutrophil-*Candida albicans* interactions. In this regard, activated HL60 cells are candicidal when infused in mice protect against systemic candidiasis *(6, 7)*. Propagation and subculturing conditions are shown in **Table 3.1**.
2. *RAW 264.7.* A number of other monocyte or macrophage cell lines are available and can be obtained from one of several

Table 3.1
Comparison of growth properties of two phagocytic cell lines

	HL-60	RRAW 264.7

Propagation	Growth property: suspension
	ATCC complete growth medium:
	  Iscove's modified Dulbecco's medium with 4 mM L-glutamine adjusted to contain 1.5 g/L sodium bicarbonate, 80%; fetal bovine serum, 20%
		Growth property: adherent
		ATCC complete growth medium:
		  Dulbecco's modified Eagle's medium with 4 mM L-glutamine adjusted to contain 1.5 g/L sodium bicarbonate and 4.5 g/L glucose, 90%; fetal bovine serum, 10%

Subculturing	Protocol: cultures can be maintained by the addition of fresh medium or replacement of medium. Interval: maintain cell density between $1 \times 10(5)$ and $1 \times 10(6)$ viable cells/mL Medium renewal: every 2–3 days	Protocol: subcultures are prepared by scraping For a 75 cm^2 flask, remove all but 10 mL culture medium. Dislodge cells from the flask substrate with a cell scraper; aspirate and add appropriate aliquots of the cell suspension into new culture vessels. Subcultivation ratio: 1:3–1:6 is recommended Medium renewal: replace or add medium every 2–3 days.

repositories. One of these, RAW264.7, can be grown in RPMI-1640 medium supplemented with 10% heat-inactivated fetal bovine serum, glutamine, and 1% streptomycin/penicillin at 37°C in 5% CO_2. Alternate conditions are defined in **Table 3.1**. At confluence of culture, cells are washed in PBS as described above and collected by gently dislodging the attached cells from the tissue culture dishes (flasks). The efficiency of collecting cells is improved by the addition of trypsin, which is then removed during subsequent washing of cells with the medium used for running the assays with yeasts or hyphae *(8)*.

3.1.2. Mouse Phagocytes

Mouse phagocytic cells can be conveniently obtained from peritoneal lavage fluids *(11)*. In the latter case, animals are injected with 1 mL of 10% proteose peptone (Difco). The peritoneum is exposed and 4 mL of sterile, ice-cold PBS contains 50 U of heparin. The fluid is collected in sterile tubes. If the lavage fluid is collected 4 h after injection of the proteose peptone, then the predominant cell is a granulocyte; and 72 h after injection, the predominant cell is a macrophage. The lavage suspension is then washed with sterile saline (600 × g, 4°C), suspended in culture medium and counted to obtain the appropriate cell density.

3.2. Preparation of Candida Blastoconidia Suspensions

1. Yeast cells of the *Candida* spp. under study are used to inoculate 5 mL of YPD liquid medium in a 50-mL tube. Cells are grown overnight at 30°C in shake culture.
2. Transfer 1 mL of the culture into a 1.5-mL tube and centrifuge for 2 min (900 × g).
3. Remove the supernatant and wash the cells with PBS twice.
4. For opsonization, incubate yeast cells in 10–50% pooled human serum for 15 min at 37°C and wash the cells twice with PBS (optional).
5. Collect yeast cells by centrifugation and resuspend in RPMI-1640 medium in an ice bath while the cell concentration is adjusted to the required number.
6. As the phagocytes are also counted, phagocytosis experiments can be set up with different effector:target (E:T) cell ratios (effector = phagocyte, target = yeasts).

3.3. Co-culture of Phagocytes with Yeast Cells

In all assays, the desired E:T ratio must be determined according to the experimental purpose. Normally, for killing assays, E:T ratios of 10:1–100:1 are chosen, but for phagocytosis assay, E:T ratios of (1:1–1:10) are appropriate.

Monolayer cell method. 96- or 24-Well culture plates can be conveniently used. To create a monolayer of phagocytes, dispense 1×10^5 cells in 75 µL into the wells (96-well plate). A yeast cell suspension of 75 µL is then added. Set up control group (phagocytes alone, or *Candida* alone). At various time points (30 min to 4 h) of incubation, the samples are processed as described in the next section (*see* **Note 1**).

Suspension sampling method. Phagocytic cells and yeasts are mixed in 14-mL round-bottom tubes (total volume 0.5–1 mL). Co-cultures are incubated at 37°C with gentle rotation for the desired period of time (30 min to 4 h). At various time points of incubation, samples are processed for microscopic observations as well as for determining the amount of organism killed (or not) by the phagocytes (see next sections).

3.3.1. Microscopic Observation

Direct light microscopic observation. For the monolayer cell method of co-culture, adherent cells can be incubated directly on a glass slide (Lab-Tek™ Chamber Slide™ System, Nalge Nunc International, USA). After 10–30 min incubation, nonphagocytized yeasts are removed by gently washing the monolayers. For the suspension sampling method of co-culture, aliquots are taken. Cell smears on slides can be made by cytospin (need special equipment), or manually, followed by flame fixation and staining of the monolayer. A useful stain for observing phagocytosis and fungal-cell interactions is the Wright-Giemsa stain. The phagocytes with or/and without internalized yeast (100 phagocytes/field, at least five fields counted) are counted. The percent phagocytosis is calculated as [*1*-(number of phagocytes ingesting yeasts/number of total phagocytes counted)].

Fluorescence assay. For FCM, yeasts are incubated in 1-mL of physiological saline containing the nonfluorescent stain, BCECF/AM (final concentration of 1 μmol L^{-1} (Roche Diagnostics, Mannheim, Germany), for 30 min at 37°C. The stain diffuses into yeasts and is subsequently cleaved by cytoplasmic esterases that generate a fluorescent, membrane impermeable product (BCECF) that has an excitation wavelength of 488 nm. The yeasts associated with phagocytes can be quantified by FCM. Further, the amount of intracellular organisms is measured using the same samples, which are then examined by epifluorescence interference contrast microscopy. Here, we focus on a description of the fluorescence microscopy method.

In this assay, one can distinguish surface-bound *Candida* from internalized yeasts *(10, 11)*. This assay is based on the observation that FITC-labeled noningested yeasts lose their green fluorescence and acquire orange fluorescence upon incubation with EtBr, which will quench the green fluorescence. The ingested yeasts retain the green fluorescence since EtBr does not enter the live host cells (*see* **Note 2**).

1. Prepare yeast cell suspension as above; 5×10^6/mL in PBS.
2. Kill yeast cells by boiling for 10 min.
3. Opsinzation (optional).
4. FITC-labeling yeast: pellet cells by centrifugation, and remove supernatant. Resuspend cells using 1 mL FITC solution, and incubate at room temperature for 30 min with periodic agitation in the dark. Wash the yeast cells three times with PBS. The cells are ready for co-culture with phagocytes.
5. After co-culture, replace the medium with PBS and add EtBr to a concentration 10 μg/mL.
6. After 1–5 min, pellet the cell mixture by centrifugation. Wash two times with PBS. Mount the cell suspension on slides and observe under fluorescence microscope. Count the phagocytes with or without ingested particles.

3.4. Killing Assays

3.4.1. Colony-Forming Unit Method

Candidicidal activity of phagocytes can be usually measured by counting survivors using colony-counting techniques (colony-forming unit (CFU)) *(12)* (*see* **Note 3**). After co-culture, aliquots of mixtures are taken out (i.e., 100 μL). Phagocytes are lysed by adding sterile water (5–10 volume) to release the internalized yeasts.

1. Serial dilutions are determined according to the original yeast number in the culture.
2. Spread 100 μL of suspension containing the intracellular yeasts on YPD plates, and count the Candida colonies after 16–24 h

of incubation at 37°C. The percent killing of *Candida* is defined as [*1* − (CFU of treatment group/CFU of control group)] × 100.

3.4.2. XTT Assays

Colorimetric assays of cellular viability are important tools in the study of eukaryotic cell activity. One of the typical methods used to measure viability is the XTT assay in which the cleavage of the tetrazolium salt, XTT, in the presence of an electron-coupling reagent, i.e., PMS, producing a soluble, colored formazan salt is measured. This conversion only occurs in viable cells *(13)*. For the XTT assay, the medium to be used must not be colored. It is also preferred to not use serum because its color may influence the absorbance values. We prefer to use RPMI-1640 medium without phenol red.

1. It is more convenient to use 96- or 24-well culture plates for culture. After the desired time interval, the reaction can be terminated by lysing the host cell (*see* **Note 4**).
2. Cells can be centrifuged to the bottom by plate centrifugation (5000 rpm, 10 min).
3. Gently remove the supernatant, and then add 150 μL water to each well of the 96-well plate.
4. Gently shake the plate for 5–10 min, centrifuge again, and remove the supernatant.
5. Add 100 μL of RPMI-1640 medium (without phenol red) into each well, 50 μL XTT/PMS solution. Incubate for 2–24 h at 37°C (you will see the color change).
6. Determine the OD value using a multiwell spectrophotometer (ELISA reader), wavelength 490 nm. The absorbance directly correlates with the cell number. The percent killing of *Candida* is defined as [*1* − (OD value of treatment group /OD value of control group)] × 100.

4. Notes

1. As *C. albicans* and other species are polymorphic, care has to be taken to ensure that germination does not occur in RPMI medium, and therefore the need for an ice bath. On the other hand, if germinated forms of the organism are needed for phagocytosis or killing assays, then the organism is grown for short times in RPMI medium at 37°C (1–2 h), then counted, and used in the same way as described above. It is advised that counting germ tubes is not always easy to do, as the germinated forms clump extensively.

2. The weakness of this assay is that dead yeast cells must be used since live yeasts concentrate FITC in their cell vacuoles and is, therefore, protected from EtBr staining. Thus, prior to using this assay, one must determine if phagocytosis of dead cells is different from that for viable cells *(14)*.

3. A disadvantage of this method more often is associated with two factors: (1) as mentioned above, the organism may have geminated, but these forms still count as a single colony even though growth has occurred. This results in a potential underestimate of CFUs. (2) Incomplete dispersal of the organisms can lead to an overestimate of microbial killing, i.e., viable cells may form small clumps that when plated on agar medium appear as single colonies.

4. PMNs are easily lysed with water; however, HL-60 cells are highly resistant to water lysis so that 0.01% SDS (final concentration) is included. Cells are treated for 10 min with this reagent followed by water to dilute/lyse the cells (SDS may interfere with XTT assay).

References

1. Lehrer, R. I., and Cline, M. J. (1969) Interaction of *Candida albicans* with human leukocytes. *J. Bacteriol.* **98**, 996–1004.
2. Cutler, J. E., and Poor, A. H. (1981) Effect of mouse phagocytes on *Candida albicans* in vivo chambers. *Infect. Immun.* **31**, 1110–1116.
3. Fradin, C., Mavor, A. L., Weindl, G., Schaller, M., Hanke, K., Kaufmann, S. H., et al. (2005) Granulocytes govern the transcriptional response, morphology, and proliferation of *Candida albicans* in human blood. *Mol. Microbiol.* **56**, 397–415.
4. Marcil, A., Harcus, D., Thomas, D. Y., and Whiteway, M. (2002) *Candida albicans* killing by RAW 264.7 mouse macrophage cells: Effects of *Candida* genotype, infection ratios, and gamma interferon treatment. *Infect. Immun.* **70**, 6319–6329.
5. Collins, S. J., Ruscetti, F. W., Gallagher, R. E., and Gallo, R. C. (1979) Normal functional characteristics of cultured human promyelocytic leukemia cells (HL60) after induction of differentiation by dimethylsulfoxide. *J. Exp. Med.* **149**, 969–974.
6. Harris, P., and Ralph, R. (1985) Human leukemic models of myelomonocytic development: A review of the HL60 and U937 cell lines. *J. Leuk. Biol.* **37**, 407–422.
7. Mullick, A., Elias, M., Harakidas, P., Marcil, A., Whiteway, M., Ge, B., et al. (2004) Gene expression in HL-60 granulocytoids and human polymorphonuclear leukocytes exposed in Candia albicans. *Infect. Immun.* **70**, 414–429.
8. Spellberg, B. J., Collins, M., French, S. W., Edwards, J. E. Jr., Fu, Y., and Ibrahim, A. S. (2005) A phagocytic cell line markedly improves survival of infected neutropenic mice. *J. Leukoc. Biol.* **78**, 338–344.
9. Vonk, A. G., Wieland, C. W., Netea M. G., and Kullberg, B. J. (2002) Phagocytosis and intracellular killing of *Candida albicans* blastoconidia by neutrophils and macrophages: A comparison of different microbiological test systems. *J. Microbiol. Methods* **49**, 55–62.
10. Peltroche-Llacsahuanga, N., Schnitzler, S., Schmidt, K., Tintelnot, R., Lutticken, R., and Hasse, G. (2000) Phagocytosis, oxidative burst, and killing of *Candida dubliniensis* and *Candida albicans* by human neutrophils. *FEMS Microbiol. Lett.* **191**, 151–155.
11. Wellington, M., Bliss, J. M., and Haidaris, C. G. (2003) Enhanced phagocytosis of *Candida* species mediated by opsonization with a recombinant human antibody single-chain variable fragment. *Infect. Immun.* **71**, 7228–7231.

12. Du, C., Calderone, R., Richert, J., and Li, D. (2005) Deletion of the *SSK1* response regulator gene in *Candida albicans* contributes to enhanced killing by human polymorphonuclear neutrophils. *Infect. Immun.* **73**, 865–871.
13. Kuhn, D. M., Balkis, Chandra J., Mukherjee, P. K., and Ghannoum M. (2003) Uses and limitations of the XTT assay in studies of *Candida* growth and metabolism. *J. Clin. Microbiol.* **41**, 506–508.
14. Schuit, K. E. (1979) Phagocytosis and intracellular killing of pathogenic yeasts by human monocytes and neutrophils. *Infect. Immun.* **24**, 932–938.

Chapter 4

Immunization Protocols for Use in Animal Models of Candidiasis

Esther Segal and Hana Sandovsky-Losica

Abstract

Immunoprotection during most forms of candidiasis (oropharyngeal, invasive) is lacking since most candidiasis patients are immunosuppressed either as a result of their allogeneic transplant, cancer chemotherapy, or HIV infection. Consequently, immunization might be considered as an unlikely way to protect patients from such infection. Nonetheless, there are a number of investigations that indicate active immunization or the passive treatment with hyperimmune, specific antibodies can result in protection in models of experimental candidiasis. The former subject, active immunization, is the subject of this chapter. We focus on recent efforts with the Als family of cell wall proteins to serve as a model, and also offer immunization methods in candidiasis models that can be adapted to any antigen of the organism.

Key words: Candidiasis, animal models, immunoprotection, Als proteins.

1. Introduction

Significant research efforts have been devoted to immunization studies to protect against candidiasis and other human fungal infections *(1–16)*. More recent efforts to develop vaccines against candidiasis have utilized purified cell wall components of expressed proteins instead of whole cell or partially purified materials as antigens *(3, 5, 17–19)*. Success of studies to raise antibodies against chosen antigens, in general, is largely dependent on several factors including: (1) the use of an antigen that will induce high titered, protective antibody, and/or cell-mediated immunity against the specific pathogen; (2) an effective immunization protocol; and (3) the availability of an appropriate model to assess the protective effect of the immunization. In starting any immunization study, it is important to bear in mind the clinical manifestations caused by *Candida* species, which in general fall into two subgroups: (1) mucosal/cutaneous and (2) deep-seated, systemic candidiasis. Consequently, the experimental models that have been developed reflect the dual nature of candidiasis and protocol efficacy should be determined in models that represent both infection types. The protocols described do not constitute an exhaustive list of the studies in the area, but rather, a selected representation of the state of art of the investigational effort in the topic. We chose to focus on a family of cell wall proteins (Als) that has been demonstrated to be important in host cell recognition, and by inference of its cell surface location, in immune cell recognition.

2. Materials

2.1. Murine Models of Invasive Candidiasis (see Refs. 8, 9, 12, 13, 15, 17–20)

1. ICR/BALB/c – immunocompetent mice – 6 week – old female mice (*See* **Note 1**).
2. ICR – immunosuppressed mice, treated with cyclophosphamide, administered intraperitoneally (i.p.) at a dose of 200 mg/kg administered at 4 days prior to immunization.
3. BALB/c ♀ × C57B/6 ♂) F_1 = (BB) $_{F1}$ tumor-bearing mice – –6- to 10-week-old female mice are inoculated i.p. with 10^4 EL-4 – lymphoma cells, one day prior to immunization.

2.2. Antigen

Materials will vary depending on the antigen under study by the investigator. In the examples given the following were used:
1. rAls1p-N (amino acids 17-432 of Als1p produced in *Saccharomyces cerevisiae*.)
2. Antigen mixed with complete Freund's adjuvant (CFA; Sigma Aldrich) injected i.p.

2.3. Immunization

1. Adjuvants. For a number of antigens, immunization may be enhanced when the antigen is suspended in an adjuvant such as incomplete Freund's adjuvant (IFA) *(8, 12)*, or in liposomes *(16)*. These are generally older methods, and currently Titermax (CytRx Corp., water in oil) is now a preferred adjuvant since responses to a specific antigen are as good as IFA, and there is less inflammation postimmunization with either of the above-mentioned adjuvants.
2. Routes of immunization, i.p. or subcutaneous (s.c.) routes are most commonly used.

2.4. Assessment of Immunization. ELISA for Mouse Antibody to rAls1pN

1. Microtiter 96-well plates.
2. rAls1p-N (5 μg/mL in PBS).
3. Tris-buffered saline (0.01 M Tris-HCl) pH 7.4; 0.15 M NaCl; containing 3% BSA.

4. Goat antimouse secondary antibody–horseradish peroxidase (Sigma), final dilution 1:5000.
5. ELISA substrate – *o*-phenylenediamine (50 mg) and 10 μL hydrogen peroxide (30% stock solution).
6. Reaction terminated with 10% H_2SO_4.

2.5. Assessment of Vaccine Efficacy

1. YPD agar plates containing 100 μg of gentamicin or other suitable media for enumeration of organism in tissues subsequent to immunization or other suitable media.

3. Methods

3.1. General Concepts

The majority of *Candida* infections effect mucosal surfaces, such as oropharyngeal or esophageal infections in HIV/AIDS patients, and these infections can be difficult to treat because of the development of drug-resistant *Candida* species. In addition, the pain associated with the lesions makes patients unwilling to eat; hence, nutrition and patient management can be compromised. Vulvovaginal candidiasis (VVC), once considered to be associated with the same underlying host response defects, e.g., HIV/AIDS, appears now to be quite different from oral mucosal disease in a number of ways, including host response. In fact, the host barriers, which are breached resulting in VVC and recurrent VVC (RVVC), are not completely understood. While invasive candidiasis (IC) comprises a smaller fraction of the total number of candidiasis infections, it is much more formidable than the mucosal/cutaneous infections in regard to morbidity, mortality, and cost to the patient. Further, therapy is delayed in many cases since a rapid, specific type of diagnosis is lacking. This type of candidiasis is the focus of research in most vaccine studies, and the rational for studying immunoprotection in IC is its severe life-threatening consequences of this disease and its at-risk consequences. Prophylactic measures to prevent IC are also predicated on its rank in the US as the third or fourth most frequent blood-borne infection.

Murine models are most commonly used in evaluating antigens for use in vaccines. However, a number of variations in the models used to date exist, i.e., different strains (e.g., ICR, BALB/c, C57Bl/6), gender, age, and state (e.g., naïve or compromised), and all may have an impact on the vaccination procedure and outcome. A common immunization protocol uses s.c. injections of the antigen. But here too, several modifications are apparent, including the number of antigen dosages, intervals between dosing or boosting, the amount of immunizing antigen, and the type of antigen, i.e., whole cell, partially or purified antigen, and the use (or not) of an adjuvant, described above.

Another common mode of immunization is i.p. inoculation, and in some cases a combination of s.c. and i.p. has been employed. Some studies involved infection by the oral route – to partially imitate the natural pathogenesis of candidiasis, originating from an endogenous source – the gastrointestinal (GI) tract.

Specifically, the protocol described below uses the *Candida albicans* r-Als (recombinant Als) antigen. The reader is directed to references on the preparation of this antigen *(18–20)*.

Assessing the efficacy of immunization (host response as well as protection) can be determined as the percentage of the immunized animals that survive the challenge in comparison to nonimmunized, infected controls. An immunization-only control is also required. Another criterion used is to determine whether the immunization prolongs the survival time of a lethal challenge, namely establishing the mean survival time (MST) of the immunized animals versus control (nonimmunized) animals. In addition to evaluation of survival, deep-organ fungal colonization as an indicator of infection can also be used. In systemic candidiasis models, most studies have used renal colonization as measurement, as the kidney constitutes a major target organ in this type of infection. Renal colonization can be demonstrated qualitatively by demonstration of fungal elements in histopathological specimens and quantitatively by enumeration of colony-forming units (CFU) in culture. Most studies use the latter as parameter.

In addition to the mouse strains described above, realize that there are numerous other inbred/outbred strains that have found application in assessing the status of immunization given as s.c. or i.p. injections. Cleansing of the skin prior to either method is necessary with alcohol preparation swabs. Sterilized needles and syringes are required, and care must be taken in maintaining sterile preparations of antigens as well as cell cultures used for immunization. Injection via the lateral tail vein (i.v.) is somewhat problematic. Again the area must be cleaned with alcohol, and no more than a 0.2-mL volume of any antigen (or cells) is used for immunization or infection. A restrainer is required in which the mouse is placed such that the tail protrudes through one end of the restrainer. Also required is a source of light that is placed beneath the tail vein to ensure its localization. Following injection, the injection site is cleansed and animals are returned to their cages. Animals should be monitored at least two to three times per day and moribund mice euthanized. In the methods described below, murine models of invasive (**Section 3.1.1**), vaginal (**Section 3.1.2**), and oral (**Section 3.1.3**) candidiasis are cited as examples in which the r-Als1p antigen is used as a vaccine. Along with each animal model, information about vaccination and efficacy of vaccination in each model is discussed. Further, in **Section 3.1.1**, the ELISA assay is described for measuring anti-rAls1 antibody.

Growth of C. albicans strains for infection. Cells are grown in YPD (1.0% yeast extract, 1.0% peptone, and 0.5% glucose (dextrose)), overnight, then standardized to a specific inoculum using a hemocytometer (see Appendix chapter). Infection is initiated either via: (a) the i.v. with either 10^4 or 10^5 viable *C. albicans*/mouse in 0.2 mL of saline; (b) i.p. inoculation with live *C. albicans* (4×10^7 or 8×10^6 cells/mouse in 0.2 mL saline or 5% mucin); (c) immunosuppressed mice – *C. albicans* given i.p. (0.2 mL 8×10^6 or 4×10^7 yeasts/mouse), or i.v. (0.2 mL, 10^4 yeasts/mouse); or (d) tumor-bearing mice – i.v. inoculation of 5×10^5 *C. albicans*/mouse. (For hints in choosing the inoculum size and strain of *C. albicans*, see **Notes 2 and 3**).

3.1.1. Recombinant Als1p-N/rAls3p-N Vaccine – IP Immunization; IV Challenge (Refs. 17–19)

1. Antigen: rAls1p-N/rAls3p-N (**Refs. 17–19**)
   Animal models: immunocompetent mice, female BALB/c both juvenile mice (8–10 weeks old), and retired breeders (≥ 6 months old) are used.
2. Immunocompromised mice are obtained by the administration of cyclophosphamide 200 mg/kg i.p. on day –2 and 100 mg/kg i.p. on day +9 relative to infection.
3. Immunization: i.p. injection of rAls1p-N mixed with CFA at day 0, boosted with another dose of the antigen with IFA at day 21.
4. Infection: immunocompetent mice are infected via the i.v. with *C. albicans* SC5314 2.5×10^5–10^6 yeast/mouse at 14 days subsequent to a booster immunization.
5. Immunocompromised mice: 2.5×10^4 blastospores/mouse. Evaluation: 28 days survival; fungal burden – CFU in kidneys, (*see* **Note 4**).
6. ELISA assays for measuring anti-rAls1-N antibody. In microtiter plates, wells are coated with 100 µL per well at 5 µg/mL of rAls1p-N in PBS. Then, mouse sera are incubated with the antigen-coated plates for 1 h at room temperature (RT), followed by washing wells with TBS containing 3% BSA (blocking step). Each well is then washed three times with TBS containing 0.05% Tween 20, followed by another three washes with TBS. Subsequently, goat antimouse secondary antibody–horse radish conjugate is added at a final dilution of 1:5000 and the plates are incubated for 1 h at RT, washed again with TBS, and substrate for the peroxidase added (*see* **Section 2**). The reaction color is allowed to develop for 30 min, after which the reaction is stopped with 10% sulfuric acid. Finally, the optical density is read at 490 nm. The ELISA titer reflects the reciprocal of the last serum dilution that gives a positive test. Controls consist of diluent only, and background levels of absorbance are subtracted from treatment samples.

3.1.2. Vaginal Candidiasis and Immunization with 1.3 rAls1p-N Vaccine–IP Immunization, Intravaginal Infection (Refs. 4, 5, 20–22)

1. Antigen: rAls1p-N/rAls3p-N.
2. Model animal: BALB/c mice are treated with estradiol valerate (30 μg administered s.c.) dissolved in peanut oil on day −3 of infection, to induce pseudoestrus.
3. Immunization: as in the systemic candidiasis model – namely, IP injection of rAls1p-N mixed with CFA at day 0, boosted with another dose of the antigen with IFA at day 21.
4. Infection: on the day of infection, the mice are first sedated by i.p. administration of 100 mg/kg ketamine. Sedated mice are then infected intravaginally with 1×10^6 blastospores of *C. albicans*.
5. Evaluation by vaginal fungal burden. At daily intervals, vaginal fluid is collected from infected animals by instilling 50 μL of PBS into the vaginal canal and then scrapping the vaginal mucosal surface. Aliquots of this suspension are then plated on standard media containing gentamicin. Enumeration of CFUs from these samples is done after a 48-h incubation of inoculated plates grown at 37°C.

3.1.3. Oropharyngeal Infection, rAls1p-N/rAls3p-N Vaccine –IP Immunization, Oral Infection (Ref. 23)

1. Antigen: rAls1p-N/ rAls3p-N
2. Model animal: mice immunocompromised by cortisone acetate (225 mg/kg s.c. on days −1, 1, and 3 relative to infection) *(23)*.
3. Immunization: as indicated in systemic and vaginal models, namely IP injection of rAls1p-N mixed with CFA at day 0, boosted with another dose of the antigen with IFA at day 21.
4. Infection: sublingually, mice are anesthetized by i.p. injection with 8 mg xylazine and 110 mg ketamine /kg. Calcium alginate urethral swabs are saturated with *C. albicans* by placing them in suspension of 10^6 organisms/mL. The saturated swabs are then placed sublingually in the oral cavity of the mice for 75 min.
5. Evaluation: the tongue and the hypoglossal tissue are excised for analysis of tissue fungal burden and histopathological analysis at 7- and 14-day postinfection.

4. Notes

1. The experimental candidiasis animal model is an important component of the system. We chose a model of nonacute systemic lethal infection, since such a model enables a more accurate follow-up for evaluation of efficacy of vaccination.
2. The establishment of a reproducible model is not easy, as it may require a delicate balance in regard to determination of the fungal inoculum (e.g., relatively small differences may exist between a fungal inoculum that leads to a lethal infection with 100% mortality rate in a short time versus an inoculum

that causes a low mortality rate). For compromised animals, lower infecting fungal doses are used.
3. The fungal inoculum may depend on the specific *C. albicans* strain used, and for different strains empirical determination may be necessary (our model was based on *C. albicans* CBS 562).
4. Mortality/survival rate and mean time to death/MST (MTD/MST) are the most objective criteria for evaluation, and are more reliable than assessment of organ colonization.

References

1. Segal, E. and Elad, D. (2004). Immunizations against fungal diseases in man and animals. In: Handbook of Fungal Biotechnology (Arora, D.K. ed.), Marcel Dekker, Inc., New-York, pp. 503–514.
2. Casadevall, A. (2002). Acquired immunity against fungi. In: Immunology of Infectious Diseases (Kaufman, S., Sher, A., Ahmed, R. eds.), ASM Press, Washington D.C., pp. 223–234.
3. Deepe, G. (2004). Preventative and therapeutic vaccines for fungal infections: from concept to implementation. *Expert Rev. Vaccines* 3, 1–9.
4. Cassone, A., De Bernardis, F. and Torososantucci, A. (2005). An outline of the role of anti-Candida antibodies in the context of passive immunization and protection from candidiasis. *Curr. Mol. Med.* 5, 377–382.
5. Magliani, W., Conti, S., Cassone, A., De Bernardis, F. and Polonelli, L. (2002) New immunotherapeutic strategies to control vaginal candidiasis. *Trends Mol. Med.* 8, 121–126.
6. Edwards, J.E. (2005) Candida species. In: Principles and Practice of Infectious Diseases (Mandell, G.L., Bennett, J.E., Dolin, R., eds.), 6th Ed, Churchill Livingstone, Philadelphia, pp. 2938–2956.
7. Segal, E. and Elad, D. (2005). Candidiasis. In: Topley and Wilson's Microbiology and Microbial Infections (Merz, W.G., Hay, R., eds.), 10th Ed, Arnold, London, pp. 579–623.
8. Levy, R., Segal, E., and Eylan, E. (1981) Protective immunity against murine candidiasis elicited by *Candida albicans* ribosomal fractions. *Infect. Immun.* 31, 874–878.
9. Segal, E. and Sandovsky-Losica, H. (1981) Experimental vaccination with *Candida albicans* ribosomes in cyclophosphamide treated animals. *Sabouraudia* 19, 267–274.
10. Levy, R., Segal, E., Eylan, E. and Barr-Nea, L. (1983) Cell-mediated immunity following experimental vaccinations with *Candida albicans* ribosomes. *Mycopathologia* 83, 161–168.
11. Levy, R., Segal, E. and Eylan, E. (1984) Detection of antibodies against *Candida albicans* ribosomes by the enzyme linked immunosorbent assay. *Mycopathologia* 87, 167–170.
12. Segal, E., Nussbaum, S. and Barr-Nea, L. (1985) Protection against systemic infections with various Candida species elicited by vaccination with *Candida albicans* ribosomes. *J. Med. Vet. Mycol.* 235, 275–285.
13. Segal, E., Sandovsky-Losica, H. and Nussbaum, S. (1985) Immune responses elicited by *Candida albicans* ribosomes in cyclophosphamide treated animals. *Mycopathologia* 89, 113–118.
14. Levy, R. and Segal, E. (1986) Induction of candicidal activity in mice by immunization with *Candida albicans* ribosomes. *FEMS Microbiol. Lett.* 36, 213–217.
15. Segal, E., Spongin, A., Levy, R. and Barr-Nea, L. (1987) Induction of protection against candidiasis in tumor bearing mice by vaccination with *Candida albicans* ribosomes. *J. Med. Vet. Mycol.* 25, 355–363.
16. Eckstein, M., Barenholtz, Y. and Segal, E. (1997) Liposomes containing *Candida albicans* ribosomes as prophylactic vaccine against disseminated candidiasis in mice. *Vaccine* 15, 220–224.
17. Ibrahim, A.S., Spellberg, B.J., Avenissian, V., Fu, Y., Filler, S.G. and Edwards, J.E., Jr. (2005) Vaccination with recombinant N-terminal domain of Als1p improves survival during murine disseminated candidiasis by

enhancing cell-mediated, not humoral, immunity. *Infect. Immun.* **73**, 999–1005.
18. Ibrahim, A.S., Spellberg, B.J., Avenissian, V., Fu, Y. and Edwards, J.E., Jr. (2006) The anti-Candida vaccine based on the recombinant N-terminal domain of Als1p is broadly active against disseminated candidiasis. *Infect. Immun.* **74**, 3039–3041.
19. Spellberg, B.J., Ibrahim, A.S., Avenasian, V., Fu, Y., Myers, C., Phan, Q.T., Filler, S.G., Yeaman, M.R. and Edwards, J.E. (2006). Efficacy of anti-Candida rAls3p-N or Alsp-N vaccines against disseminated and mucosal candidiasis. *J. Infect. Dis.* **194**, 256–260.
20. De Bernardis, F., Lorenzini, R. and Cassone, A. (1999). Rat model of Candida vaginal infection. In: Handbook of Animal Models in Infection (Zak, O, Sande, M.A. eds.), Academic Press, New York, pp. 735–740.
21. Cardenas-Freytag, L., Steele, C., Wormley, F.L., Cheng, E., Clements, J.R. and Fidel, J.R. (2002). Partial protection against experimental candidiasis after mucosal vaccination with heat killed Candida albicans and mucosal adjuvant LT (R192G). *Med. Mycol.* **40**, 291–299.
22. De Bernardis, F., Boccanera, M., Adriani, D., Girolamo, A. and Cassone, A. (2002). Intravaginal and intranasal immunizations are equally effective in inducing vaginal antibodies and conferring protection against vaginal candidiasis. *Infect. Immun.* **70**, 2725–2729.
23. Segal, E., Baranetz, T., Sandovsky-Losica, H., Gov, Y., Teicher, S. and Dayan, D. (1999). Experimental oral murine candidiasis and attempts of prevention. *J. Med. Mycol.* **9**, 10–15.

Part II
Virulence and Biofilms

Chapter 5

Penetration of Antifungal Agents Through *Candida* Biofilms

L. Julia Douglas

Abstract

A filter disk assay is described, which measures the penetration of antifungal agents through *Candida* biofilms. The technique involves forming a colony biofilm on a polycarbonate membrane filter, and capping it with a second, smaller membrane filter followed by a wetted paper disk of the type used in zone-of-inhibition assays. The entire assembly is transferred to agar medium containing the antifungal agent of interest. During subsequent incubation, the drug diffuses out of the agar and through the biofilm 'sandwich' to the moistened paper disk. The drug concentration in the disk can be determined by measuring the zone of growth inhibition that it produces on medium seeded with an indicator strain of *Candida albicans* in standard bioassays. Additional procedures are outlined for determining the viabilities of drug-treated biofilms and for examining biofilm morphology by scanning electron microscopy.

Key words: Biofilm, Candida, antifungal agent, drug penetration, scanning electron microscopy.

1. Introduction

During recent years, *Candida albicans*, together with related *Candida* species, has become one of the commonest agents of hospital-acquired infections (1, 2). Many of these are implant-associated infections, in which adherent fungal populations, or biofilms, are formed on the surfaces of medical devices including catheters, prosthetic heart valves and endotracheal tubes (3, 4). Biofilm cells are organized into structured communities enclosed within a matrix of extracellular material. They are phenotypically distinct from planktonic (suspended) cells; most notably, they are significantly less susceptible to antifungal agents (5–15). This property makes implant infections difficult to treat and often the implant must be removed.

The mechanisms of biofilm resistance to antimicrobial agents are not fully understood. One long-standing hypothesis is that the biofilm matrix resists drug penetration by forming a reaction–diffusion barrier *(16)*, and that only the surface layers are exposed to a lethal dose of the agent. The extent to which the matrix acts as a barrier to drug diffusion would depend on the chemical nature of both the antimicrobial agent and the matrix material. This protocol allows the penetration of antifungal agents through *Candida* biofilms to be investigated *(17)* using a filter disk assay modified from one previously reported for bacterial biofilms *(18)*. The procedure is also suitable for measuring drug penetration through mixed-species biofilms *(17)*. In addition, methods for determining the viabilities of drug-treated biofilm cells, and for examining the biofilms by scanning electron microscopy, are described.

2. Materials

2.1. Preparation of Biofilm Inoculum

1. Slope cultures of *Candida* species on Sabouraud dextrose agar (Difco).
2. Yeast nitrogen base (YNB) medium (Difco) containing 50 mM glucose: batches of 50 mL of medium in 250-mL Erlenmeyer flasks.
3. Wash buffer: 0.15 M phosphate-buffered saline (PBS; pH 7.2).

2.2. Biofilm Formation

1. Polycarbonate membrane filters (Whatman): diameter 25 mm and pore size 0.2 µm. Sterilize by exposure to UV radiation for 15 min on both sides.
2. YNB agar containing 50 mM glucose in Petri dishes.

2.3. Antifungal Agents

1. Flucytosine (Sigma), fluconazole (Pfizer) and caspofungin (Merck) solutions in sterile distilled water, prepared immediately before use.
2. Voriconazole (Pfizer) and amphotericin B (Sigma) solutions in dimethyl sulfoxide, prepared immediately before use. Other antifungal agents may be used, dissolved fresh in the appropriate solvent.

2.4. Penetration of Biofilms by Antifungal Agents

1. Antifungal-agent-supplemented agar: using a sterile filtration unit (Sartorius), filter the drug solution into culture medium (YNB agar containing 50 mM glucose) buffered with 0.165 M morpholinepropanesulfonic acid (Sigma) to pH 7, and kept molten at 50°C. High drug concentrations are used (*see* **Note 1**).
2. Polycarbonate membrane filters (Whatman): diameter 13 mm and pore size 0.2 µm. Sterilize by exposure to UV radiation for 15 min on both sides.

3. Paper concentration disks (Becton Dickinson): diameter 6 mm. Sterilize by exposure to UV radiation for 15 min on both sides.
4. Plates of YNB agar containing 200 mM glucose and seeded with 150 µL of a standardized suspension of planktonic *C. albicans* (used here as an indicator organism and adjusted to an optical density at 520 nm of 1.0).

2.5. Viable Counts of Biofilm Cells Exposed to Antifungal Agents

1. Plates of YNB agar containing 200 mM glucose.
2. Dilution buffer: 0.15 M PBS (pH 7.2).

2.6. Scanning Electron Microscopy

1. 2.5% (v/v) Glutaraldehyde in PBS.
2. 1% (w/v) Osmium tetroxide.
3. 1% (w/v) Uranyl acetate.
4. Series of ethanol solutions ranging, in 10% increments, from 30% (v/v) ethanol in distilled water to dried absolute ethanol.

3. Methods

3.1. Preparation of Biofilm Inoculum

1. Batches of YNB medium containing 50 mM glucose (50 mL in 250-mL Erlenmeyer flasks) are inoculated from fresh *Candida* slopes and incubated at 37°C for 24 h in an orbital shaker operating at 60 rpm.
2. Cells are harvested and washed twice in PBS.
3. Before use in biofilm experiments, all washed cell suspensions are adjusted to an optical density at 600 nm of 0.2.
4. If mixed-species biofilms are to be tested (*see* **Note 2**), equal volumes of the standardized suspension of each organism are mixed immediately before use.

3.2. Biofilm Formation

1. Biofilms are grown on membrane filters resting an agar culture medium in Petri dishes. Sterile polycarbonate membrane filters (diameter 25 mm) are placed on the surface of YNB agar containing 50 mM glucose (*see* **Note 3**).
2. A standardized cell suspension (50 µL) is applied to the surface of each membrane and the plates are incubated at 37°C.
3. After 24 h, the membrane-supported biofilms are transferred to fresh agar for a further 24 h, giving a total incubation time of 48 h for biofilm formation.

3.3. Penetration of Biofilms by Antifungal Agents

1. After biofilm formation on membrane filters, smaller sterile polycarbonate membrane filters (diameter 13 mm) are carefully placed on top of the 48-h-old biofilms.

2. Sterile paper concentration disks (diameter 6 mm) are moistened with growth medium (normally 29 µL; see **Note 4**) and positioned on top of the 13-mm-diameter membranes.
3. Biofilms sandwiched between the membranes and moistened disks are transferred to plates of antifungal-agent-containing medium using sterile forceps and incubated at 37°C for time periods ranging from 60 min to 6 h (see **Fig. 5.1** and **Note 5**).
4. The amount of antifungal agent, which penetrates each biofilm and reaches the concentration disk, is determined by using the disk in a standard drug diffusion assay. After the appropriate exposure time, concentration disks are removed from the biofilm 'sandwiches' and placed on plates of YNB agar containing 200 mM glucose (see **Note 6**), which have been seeded with the indicator strain of *C. albicans*. The plates are incubated at 37°C for 24 h.
5. Zones of growth inhibition are measured and used to determine the concentration of active antifungal agent in the disks by the reference to a standard curve prepared by using drug solutions of different concentrations but fixed volumes (see **Note 7**).
6. Control assays are included where concentration disks are placed on the two-membrane system to which no cells have been added, i.e., the unit without the biofilm.
7. The drug concentration that penetrates the biofilm (C) is divided by the drug concentration determined for the control (C_0) to provide a normalized penetration curve (see **Figs. 5.2** and **5.3**, and **Note 8**).

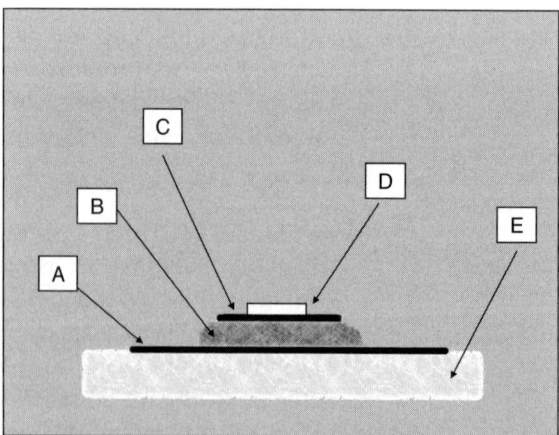

Fig. 5.1. Experimental system used to determine the penetration of antifungal agents through biofilms. The biofilm (*B*) is formed on a 25-mm-diameter membrane filter (*A*) resting on YNB agar. A second, smaller filter (*C*) is placed on *top* of the biofilm, and a moistened concentration disk (*D*) is positioned on *top* of the second filter. The entire assembly is then transferred to antifungal-containing agar (*E*).

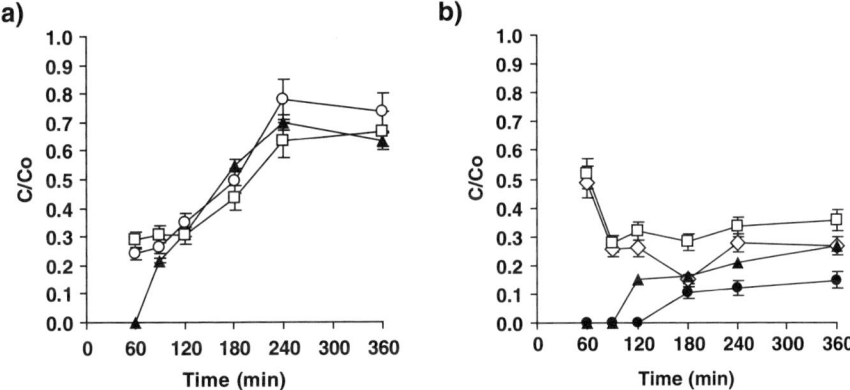

Fig. 5.2. Penetration of flucytosine through biofilms of (**a**) *C. albicans* GDH 2346 (*closed triangles*), *C. albicans* GDH 2023 (*open squares*), and *C. albicans* GRI 682 (*open circles*); (**b**) *C. krusei* (*open squares*), *C. glabrata* (*open diamonds*), *C. parapsilosis* (*closed triangles*) and *C. tropicalis* (*closed circles*). Error bars indicate the standard errors of the means. C, drug concentration on distal edge of biofilm; C_o, drug concentration measured in control. Reproduced from **Ref.** *(17)* with permission from the American Society of Microbiology.

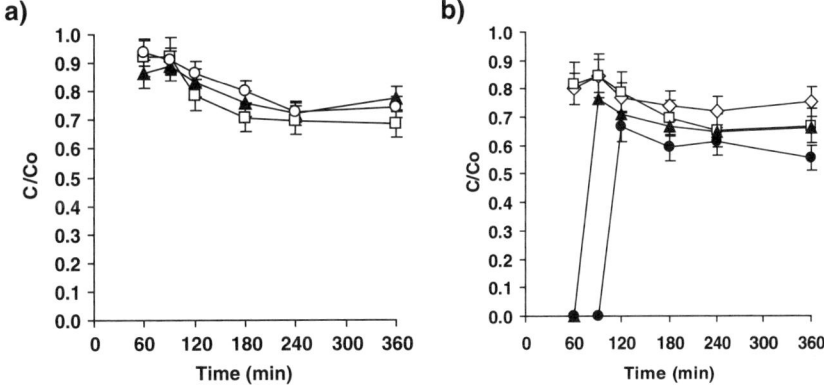

Fig. 5.3. Penetration of fluconazole through biofilms of (**a**) *C. albicans* strain GDH 2346 (*closed triangles*), *C. albicans* GDH 2023 (*open squares*), and *C. albicans* GRI 682 (*open circles*); (**b**) *C. krusei* (*open squares*), *C. glabrata* (*open diamonds*), *C. parapsilosis* (*closed triangles*) and *C. tropicalis* (*closed circles*). Error bars indicate the standard errors of the means. C, drug concentration on distal edge of biofilm; C_o, drug concentration measured in control. Reproduced from **Ref.** *(17)* with permission from the American Society of Microbiology.

3.4. Viable Counts of Biofilm Cells Exposed to Antifungal Agents

1. After incubation on antifungal-agent-containing agar, biofilm cells are gently scraped from the membranes with a sterile scalpel and resuspended in 10 mL of PBS.
2. Serial dilutions (10^{-1}–10^{-6}) of each biofilm cell suspension are prepared in PBS.
3. Triplicate samples (0.1 mL) of the 10^{-4}, 10^{-5} and 10^{-6} dilutions are spread on YNB agar containing 200 mM glucose and these plates are incubated at 37°C for 24 h.
4. In control assays, membranes are transferred to growth medium containing no antifungal agent and viable counts determined as above.

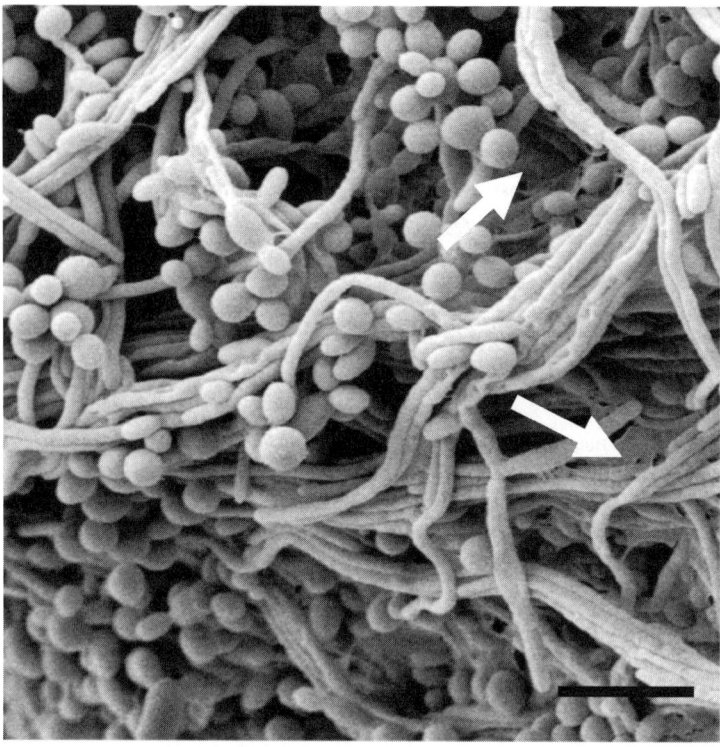

Fig. 5.4. Scanning electron micrograph of a 48-h membrane-supported biofilm of *C. tropicalis*. *Arrows* indicate extracellular matrix material. *Bar*, 10 μm. Reproduced from **Ref.** *(17)* with permission from the American Society of Microbiology.

3.5. Scanning Electron Microscopy

1. Biofilms formed on polycarbonate membranes are fixed in 2.5% (v/v) glutaraldehyde in PBS for 1 h at room temperature and immersed in 1% (w/v) osmium tetroxide for 1 h. They are then washed three times with distilled water (3 mL), treated with 1% (w/v) uranyl acetate for 1 h and washed again with distilled water (3 mL).
2. Dehydration of the samples is achieved by immersion in a series of ethanol solutions ranging, in 10% increments, from 30% (v/v) ethanol in distilled water to dried absolute ethanol.
3. The final step before gold coating is drying. Simple air-drying in a desiccator will suffice for many purposes, although it can lead to a significant loss of the biofilm extracellular matrix. A typical electron micrograph of a membrane-supported biofilm is shown in **Fig. 5.4**.

4. Notes

1. Drug concentrations are selected on the basis of their ability to give large zones of growth inhibition in control assays for drug penetration. Examples of recommended concentrations would

be flucytosine, 30 times the MIC for planktonic cells of the same strain; fluconazole, 60 times the MIC; voriconazole, 220 times the MIC; and amphotericin B, 60 times the MIC.

2. The protocol can also be used to test drug penetration through mixed-species biofilms. In such cases, a medium should be selected that is best able to support the growth of both the species. For example, tryptic soy broth (Difco) supports the growth of both *Candida* species and *Staphylococcus* species, and is a suitable choice for propagation of biofilms containing both the organisms.

3. If mixed-species biofilms are to be tested, a suitable solid medium should be selected. Tryptic soy agar is an excellent medium for *Candida/Staphylococcus* biofilms.

4. Because of an occasional variation in disk thickness, a slightly lower or higher volume of medium is sometimes required to saturate the disks. Wetting of the disks helps prevent the capillary action of the antifungal medium through the biofilms.

5. Exposure times of 60, 90, 120, 180, 240 and 360 min are convenient for most purposes.

6. YNB agar containing 200 mM glucose, rather than 50 mM, gives optimal lawn production by the indicator strain.

7. A standard curve is obtained by plotting the log of the drug concentration used against the diameter of the zone of growth inhibition.

8. Colony biofilms formed on membrane filters appear to lack the water channels that typically surround matrix-enclosed microcolonies in many other biofilms. The possibility that drugs simply move through water channels without reaching cells deep within the microcolonies is, therefore, largely eliminated by using this model system.

References

1. Calderone, R. A. (ed.) (2002) *Candida and Candidiasis*. ASM Press, Washington, D.C.
2. Kojic, E. M., and Darouiche, R. O. (2004) *Candida* infections of medical devices. *Clin. Microbiol. Rev.* **17**, 255–267.
3. Douglas, L. J. (2003) *Candida* biofilms and their role in infection. *Trends Microbiol.* **11**, 30–36.
4. Donlan, R. M. (2001) Biofilms and device-associated infections. *Emerg. Infect. Dis.* **7**, 277–281.
5. Hawser, S. P., and Douglas, L. J. (1995) Resistance of *Candida albicans* biofilms to antifungal agents in vitro. *Antimicrob. Agents Chemother.* **39**, 2128–2131.
6. Baillie, G. S., and Douglas, L. J. (1998) Effect of growth rate on resistance of *Candida albicans* biofilms to antifungal agents. *Antimicrob. Agents Chemother.* **42**, 1900–1905.
7. Baillie, G. S., and Douglas, L. J. (1998) Iron-limited biofilms of *Candida albicans* and their susceptibility to amphotericin B. *Antimicrob. Agents Chemother.* **42**, 2146–2149.
8. Baillie, G. S., and Douglas, L. J. (1999) *Candida* biofilms and their susceptibility to antifungal agents. *Meth. Enzymol.* **310**, 644–656.
9. Baillie, G. S., and Douglas, L. J. (1999) Role of dimorphism in the development of *Candida albicans* biofilms. *J. Med. Microbiol.* **48**, 671–679.
10. Chandra, J., Mukherjee, P. K., Leidich, S. D., Faddoul, F. F., Hoyer, L. L., Douglas, L. J.,

and Ghannoum, M. A. (2001) Antifungal resistance of candidal biofilms formed on denture acrylic in vitro. *J. Dent. Res.* **80**, 903–908.
11. Chandra J., Kuhn, D. M., Mukherjee, P. K., Hoyer, L. L., McCormick, T., and Ghannoum, M. A. (2001) Biofilm formation by the fungal pathogen *Candida albicans*: development, architecture, and drug resistance. *J. Bacteriol.* **183**, 5385–5394.
12. Lewis, R. E., Kontoyiannis, D. P., Darouiche, R. O., Raad, I. I., and Prince, R. A. (2002) Antifungal activity of amphotericin B, fluconazole, and voriconazole in an in vitro model of *Candida* catheter-related bloodstream infection. *Antimicrob. Agents Chemother.* **46**, 3499–3505.
13. Ramage, G., VandeWalle, K., Wickes, B. L., and Lopez-Ribot, J. L. (2001) Standardized method for in vitro antifungal susceptibility testing of *Candida albicans* biofilms. *Antimicrob. Agents Chemother.* **45**, 2475–2479.
14. Ramage, G., VandeWalle, K., Wickes, B. L., and Lopez-Ribot, J. L. (2001) Biofilm formation by *Candida dubliniensis*. *J. Clin. Microbiol.* **39**, 3234–3240.
15. Andes, D., Nett, J., Oschel, P., Albrecht, R., Marchillo, K., and Pitula, A. (2004) Development and characterization of an in vivo central venous catheter *Candida albicans* biofilm model. *Infect. Immun.* **72**, 6023–6031.
16. Gilbert, P., Maira-Litran, T., McBain, A. J., Rickard, A. H., and Whyte, F. W. (2002) The physiology and collective recalcitrance of microbial biofilm communities. *Adv. Microb. Physiol.* **46**, 203–256.
17. Al-Fattani, M. A., and Douglas, L. J. (2004) Penetration of *Candida* biofilms by antifungal agents. *Antimicrob. Agents Chemother.* **48**, 3291–3297.
18. Anderl, J. N., Franklin, M. J., and Stewart, P. S. (2000) Role of antibiotic penetration limitation in *Klebsiella pneumoniae* biofilm resistance to ampicillin and ciprofloxacin. *Antimicrob. Agents Chemother.* **44**, 1818–1824.

Chapter 6

Candida Biofilm Analysis in the Artificial Throat Using FISH

Bastiaan P. Krom, Kevin Buijssen, Henk J. Busscher, and Henny C. van der Mei

Abstract

Biofilm formation is a common complication of the use of prosthetic devices. In clinical settings, biofilms can be comprised of one or more microbial species. In order to investigate the interaction between different species within a biofilm, a reproducible, reliable model system has to be utilized and an appropriate system for species identification applied. The present chapter describes the artificial throat model, a model system for growing mixed species biofilms on shunt prostheses. The model is used in conjugation with fluorescent *in situ* hybridization (FISH), which facilitates identification and localization of the resident microorganisms within biofilms.

Key words: Artificial throat, shunt prosthesis, mixed-species biofilm, microbial interactions, fluorescent *in situ* hybridization, FISH.

1. Introduction

1.1. Biofilms and Infection

Many microbial infections can be linked to the ability of the infectious agents to establish biofilms (1), and thus, biofilm formation is of concern in clinical settings. More recently, attention has been given to the study of fungal biofilms, most notably those formed by *Candida albicans* (2, 3), but significantly less information is available concerning their development. Biofilms, in a clinical setting, are organized microbial communities that can form at bodily sites of infection and on the surfaces of prosthetic devices, most notably to the topic of this chapter, shunt prostheses (4). Biofilm formation in general proceeds in distinct stages:
1. Initial adhesion of microorganisms to the substratum.
2. Irreversible attachment of microorganisms to the substratum.
3. Growth and proliferation of the biofilm.
4. Maturation, including extracellular polymeric substance (EPS) formation and cell detachment.

With regard to *C. albicans*, a number of model systems have been developed to analyze the stages of biofilm development, and in addition, to determine unique characteristics of *C. albicans* biofilms. Investigations have included studies of biofilm development on clinically relevant materials such as catheter discs *(5, 6)*, denture acrylics *(7)*, as well as in microtiter plate wells *(8–10)* and (micro)fermenter systems *(11, 12)*. Importantly, an artificial throat model has been developed in Groningen *(13)* to study biofilm formation on voice prostheses, and this model for mixed (bacterial – *C. albicans*) biofilms is described in this chapter.

1.2. Laryngeal Cancer and Shunt Prostheses

Laryngeal and pharyngeal cancers are the most frequently occurring cancers in the upper airway and digestive tract. Approximately 5–20% of all patients will require total laryngectomy. The most disabling consequence of this procedure is generally considered to be loss of vocal function *(14)*. The insertion of a silicone rubber shunt prosthesis (sometimes referred to as voice prosthesis) in a surgically created tracheoesophageal fistula was a major step forward in the speech rehabilitation of laryngectomized patients, and is now generally considered to be superior to any other form of substitute voice production, such as esophageal or electrolaryngeal speech.

Several types of shunt prostheses have been developed since 1980, most of which are made of medical-grade silicone rubber because of its excellent mechanical and molding properties. Unfortunately, silicone rubber is also an excellent substratum for microbial biofilm formation, especially by mixed bacterial *Candida* species, with *C. albicans* being most prevalent *(15–18)*. In general, biofilm growth on the shunt prostheses causes increased airflow resistance or leakage through or around the device. This necessitates the removal of the device on average every 16 weeks *(19)*, but in some patients replacement is needed within 2 weeks after implantation *(20)*. In addition, *Candida* species have been described to be most responsible for failure of the device due to their ability to grow into the silicone rubber and deteriorate its mechanical properties *(17–20)*.

1.3. The Artificial Throat Model

The artificial throat model (**Fig. 6.1**) is basically a modified Robbins device fitted with shunt prostheses, inserted in a temperature-controlled, heat sterilizable incubation chamber, in which the environment of a human throat can be mimicked (*See* **Note 1**). The device is linked to a roller pump that is linked through a series of valves to three reservoirs containing either the inoculum, phosphate-buffered saline (PBS), or an appropriate growth medium. The effluent is collected and can be analyzed for biological or chemical composition. In addition to the use of the Groningen® shunt prostheses, this system is also compatible with other commonly used shunt prostheses, like Provox® and Voice Master® *(13)*.

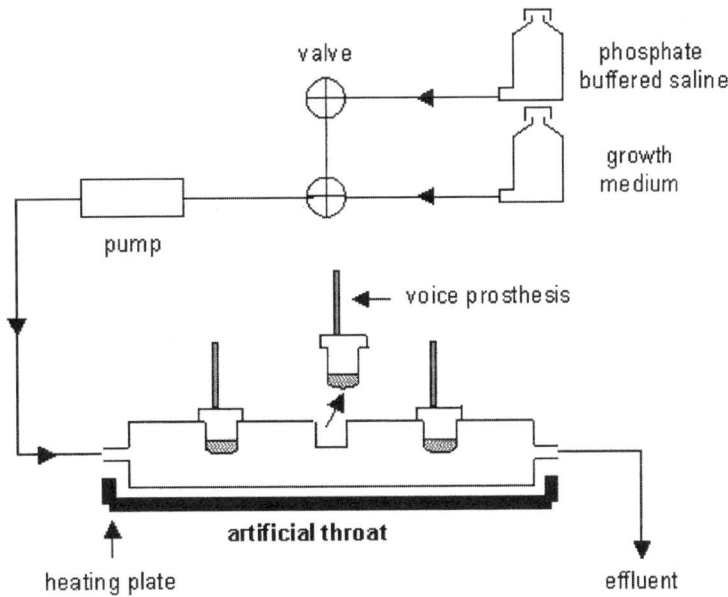

Fig. 6.1. Schematic representation of the modified Robbins device used as artificial throat. (From **Ref.** *(25)* with permission).

1.4. Fluorescent In Situ *Hybridization*

Fluorescent *in situ* hybridization (FISH) uses specific probes to identify species, or groups of microorganisms in complex microbial communities *(21)*. In combination with confocal laser scanning microscopy (CLSM), it can be applied to biofilm research if care is taken not to disturb the biofilm architecture during fixation and hybridization steps *(22, 23)*. Furthermore, FISH is an established method for determination of spatial arrangement of species in a biofilm. In our laboratory in Groningen, FISH has been applied to investigate the composition of biofilms on explanted voice prostheses *(24)* and on biofilms grown in the artificial throat model *(25)*.

2. Materials

2.1. The Artificial Throat

1. 30% Brain heart infusion, 70% defined yeast medium: 11.1 g/L BHI OXOID, Basingstoke, UK, 7.5 g/L glucose, 3.5 g/L $(NH_4)_2SO_4$, 1.5 g/L L-asparagine, 10 mg/L L-histidine, 20 mg/L DL-methionine, 20 mg/L DL-tryptophane, 1 g/L KH_2PO_4, 500 mg/L $MgSO_4 \cdot 7H_2O$, 500 mg/L NaCl, 500 mg/L $CaCl_2 \cdot 2H_2O$, 100 mg/L yeast extract, 500 µg/L H_3BO_3, 400 µg/L $ZnSO_4 \cdot 7H_2O$, 120 µg/L $Fe(III)Cl_3$,

200 μg/L Na$_2$MoO$_4$·2H$_2$O, 100 μg/L KI, and 40 μg/L CuSO$_4$·5H$_2$O (Merck Darmstadt, Germany).
2. PBS: 10 mM potassium phosphate, 150 mM NaCl, pH 7.0.
3. Modified Robbins devices (*see* **Note 1**).
4. Heating plate.
5. Shunt prostheses:
   a. Groningen®
   b. Provox®
   c. Voice Master®

2.2. Strains

1. The initial inoculum that is introduced to the shunt prosthesis must be selected on the basis of good biofilm formation from the total cultivable microflora from explanted shunt prostheses *(13, 26, 27)*, and the species involved are listed in **Table 6.1** (*see* **Note 2**).

2.3. Sample Preparation

1. 4% Paraformaldehyde solution (MP Biomedicals, LLC, Eschwege, Germany) in PBS pH 7.2–7.5.
2. Ethanol:PBS (1:1).
3. Using a surgical blade, cross-sections of the valve portion of the prosthesis are cut and placed on a microscope slide.

2.4. Probes

1. Probes are obtained from Eurogentec (Seraing, Belgium).
2. Probe EUB338 *(28)* is specific for the domain bacteria (labeled with rhodamine providing a red signal) and probe EUK516 *(28)* is specific for the domain eukarya (labeled with fluorescein-isothyocyanate (FITC) providing a green signal).

2.5. Fluorescent In Situ Hybridization

1. Hybridization buffer: 0.9 M NaCl, 20 mM Tris/HCl pH 7.2, and 0.01% SDS, is prewarmed to 50°C.
2. Washing buffer: 0.9 M NaCl, 20 mM Tris/HCl pH 7.2, is prewarmed to 50°C.
3. A moist dark incubation chamber (*see* **Note 3**).

Table 6.1
The microbial species used in the artificial throat model

Yeast species	Bacterial species
Candida albicans	Streptococcus salivarius
C. tropicalis	Rothia dentocariosa
	Staphylococcus aureus
	S. epidermidis

2.6. Confocal Laser Scanning Microscopy

1. CLSM model LEICA TCS SP2 (Leica Microsystems, Heidelberg GmbH, Heidelberg, Germany)
2. Lasers: UV laser, He-Ne laser, and Ar laser
   Wavelength: excitation 400 nm emission 490 (Calcofluor)
   excitation 488 nm emission 514 (FITC)
   excitation 543 emission 580 (Rhodamine)
3. Slime stain: 0.2 µg/mL Calcofluor White (Sigma, Steinheim, Germany)
4. Mounting medium Vectashield (Vector Laboratories, Inc. Burlingame CA, USA).

3. Methods

3.1. The Artificial Throat

1. Prostheses are placed in the sterilized chamber and sterile tubing is connected.
2. Overnight cultures of microorganisms to be used (**see Table 6.1**) are mixed.
3. Biofilm formation is initiated by adherence of microorganisms to the surface of inserted shunt prostheses. Adherence is achieved by static incubation of the chamber with the microbial cell suspension for 5 h at 36–37°C.
4. After initial adherence in the chamber, the artificial throat is filled with 30% brain heart infusion 70% defined yeast medium, and biofilms are allowed to develop by static incubation at 36–37°C. After 3 days of initial biofilm formation, the medium is removed from the artificial throat by flowing PBS (pH 7) through the system.
5. To mimic the environmental conditions in the human throat, the biofilms are left in the moist environment of the drained device.
6. The device is perfused with PBS and drained, three times each day.
7. At the end of each day, the chamber is filled with growth medium, 30% brain heart infusion 70% defined yeast medium, for 30 min after which it is drained and left in the moist environment overnight.
8. After 12 days of biofilm formation, biofilms formed by this technique are highly comparable to those found on shunt prostheses from laryngectomized patients in both microbial composition and microscopic appearance *(13)*.

3.2. Sample Preparation

1. Prostheses are removed from the artificial throat and placed in PBS, pH 7 at 4°C.
2. Samples are fixed with 4% paraformaldehyde for 24 h at 4°C.
3. Samples are stored in ethanol:PBS (1:1) at 4°C (or –20°C for long-term storage).

4. Using a surgical blade, cross-sections of the valve section of the prosthesis are cut and placed on a microscope slide.

3.3. Fluorescent In Situ Hybridization

1. Hybridization buffer is prewarmed to 50°C.
2. Both probes (EUB338 and EUK516) are added to the buffer to a final concentration of 5 ng/μL.
3. Hybridization buffer of 50 μL is applied to each sample (*see* **Note 4**).
4. A moist dark chamber is obtained by placing wetted towels in a dark microscope slides holder.
5. The slides are hybridized at 50°C for 17 h in the moist dark chamber.
6. The unbound probe is removed by washing once with prewarmed washing buffer.

3.4. Confocal Laser Scanning Microscopy

1. Before microscopic analysis, the slides are rinsed with ultrapure water.
2. Slides are air-dried and prepared for microscopy (*see* **Note 5**).
3. An example of expected results is shown in **Figs. 6.2A and B**.

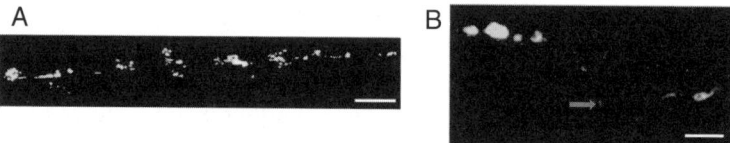

Fig. 6.2. **Panel A** and **B**. FISH analysis on the shunt prostheses from the artificial throat. (From **Ref.** *(25)* with permission).

4. Notes

1. The modified Robbins device has been made from Perspex plastic or stainless steel (**Fig. 6.3A**). The latter is our preferred system because of its durable, heat sterilizable nature. The devices can be fitted with 3 or 10 shunt prosthesis for small- or large-scale studies (**Fig. 6.3B**). The devices are placed upside down on a heating plate allowing full heating of the whole chamber (**Fig. 6.3C**). To obtain the best results, the whole device is covered with cotton and aluminum foil to maintain stable temperatures.
2. In the artificial throat model, multiple bacterial and fungal species can be combined. The selection used here represents those microorganisms that are commonly encountered on explanted voice prostheses with short life-times.
3. This can be achieved by placing wetted towels in a dark microscopy slide holder (Kartell, Italy).

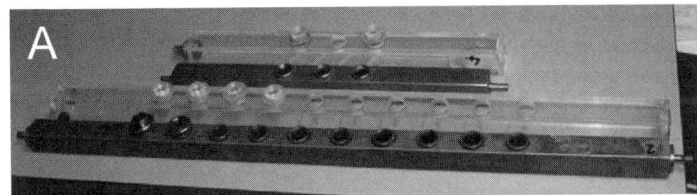

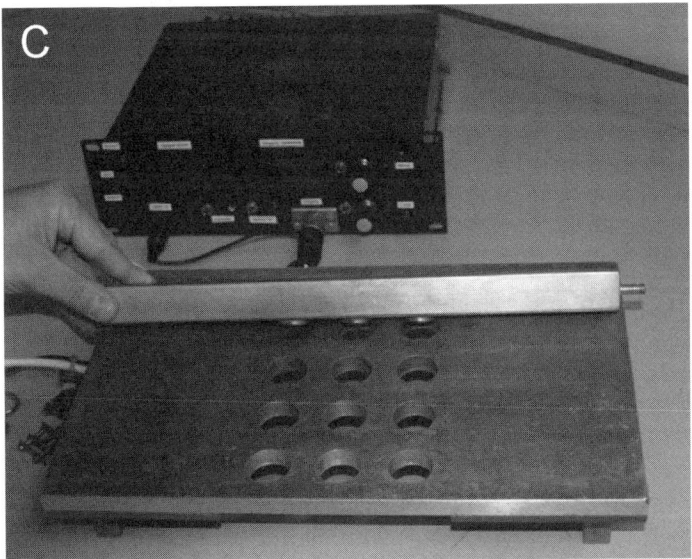

Fig. 6.3. Elements of the artificial throat. **Panel A**. Four different Robbins devices made of Perspex plastic or stainless steel with either 3 or 10 positions for shunt insertion. **Panel B** shows insertion positions for shunt prostheses (*right*), the fixation devices for these shunt prostheses (*middle*), and the pistons used to close the pistons during sterilization or when the piston is kept empty during an experiment. **Panel C** shows positioning of the Robbins device on the heating plate (in this case, the device with 3 positions is shown; for the device with 10 positions, larger heating plates are used).

4. To get fixation of the sample to the slide, a layer of silicone grease can be applied (**Fig. 6.4A**). In order to submerge the sample in the hybridization buffer, plastic rings were used around the sample preventing the solution from drifting away (**Fig. 6.4B**).

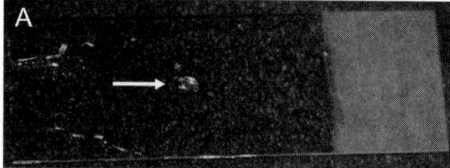 

Fig. 6.4. Placement of the sample (*arrow*) on a microscope slide covered with silicon grease for fixation purposes (**Panel A**). **Panel B** shows placement of a plastic O-ring to create a small hybridization chamber around the sample. This chamber fits 50 µL of hybridization fluid to minimize probe usage.

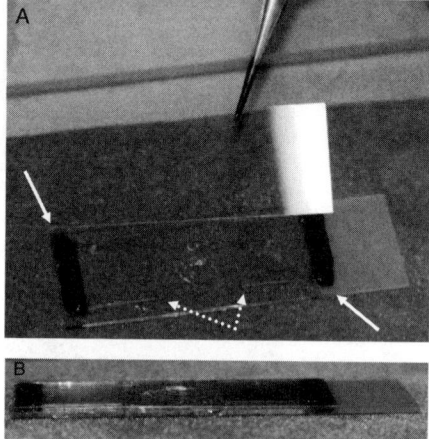

Fig. 6.5. Building a coverslip bridge in order to prevent sample compression during microscopy. The *arrows* indicate small coverslips that have been glued to the microscope slide using nail polish. Depending on the sample thickness, a number of smaller coverslips can be stacked and finally covered with a large coverslip that is also glued to the slide using nail polish (**Panel B**).

5. Depending on the thickness of the cross-sections, coverslip "bridges" can be made to cover the sample (**Figs. 6.5A** and **B**) and prevent compression during microscopy. This is of critical nature due to the structural heterogeneity of biofilms.

Acknowledgments

The authors acknowledge the valuable technical assistance of Jelly Atema-Smit and helpful suggestions of Hermie J.M. Harmsen.

References

1. Potera, C. (1999) Forging a link between biofilms and disease. *Science* **283**, 1837–1839.
2. Douglas, L. J. (2002) Medical importance of biofilms in *Candida* infections. *Rev. Iberoam. Micol.* **19**, 139–143.
3. Douglas, L. J. (2003) *Candida* biofilms and their role in infection. *Trends Microbiol.* **11**, 30–36.
4. Costerton, J. W., Lewandowski, Z., Caldwell, D. E., Korber, D. R., and Lappin-Scott, H.

M. (1995) Microbial biofilms. *Annu. Rev. Microbiol.* **49**, 711–745.

5. Hawser, S. P., and Douglas, L. J. (1994) Biofilm formation by *Candida* species on the surface of catheter materials in vitro. *Infect. Immun.* **62**, 915–921.

6. Hawser, S. P., and Douglas, L. J. (1995). Resistance of *Candida albicans* biofilms to antifungal agents in vitro. *Antimicrob. Agents Chemother.* **39**, 2128–2131.

7. Chandra, J., Kuhn, D. M., Mukherjee, P. K., Hoyer, L. L., McCormick, T., and Ghannoum, M. A. (2001) Biofilm formation by the fungal pathogen *Candida albicans*: development, architecture, and drug resistance. *J. Bacteriol.* **183**, 5385–5394.

8. Ramage, G., Vande Walle, K., Wickes, B. L., and Lopez-Ribot, J. L. (2001) Standardized method for in vitro antifungal susceptibility testing of *Candida albicans* biofilms. *Antimicrob. Agents Chemother.* **45**, 2475–2479.

9. Ramage, G., Vandewalle, K., Bachmann, S. P., Wickes, B. L., and Lopez-Ribot, J. L. 2002. In vitro pharmacodynamic properties of three antifungal agents against preformed *Candida albicans* biofilms determined by time-kill studies. *Antimicrob. Agents Chemother.* **46**, 3634–3636.

10. Bachmann, S. P., Ramage, G., Vandewalle, K., Patterson, T. F., Wickes, B. L., and Lopez-Ribot, J. L. (2003) Antifungal combinations against *Candida albicans* biofilms in vitro. *Antimicrob. Agents Chemother.* **47**, 3657–3659.

11. Lewis, R. E., Kontoyiannis, D. P., Darouiche, R. O., Raad, I. I., and Prince, R. A. (2002) Antifungal activity of amphotericin B, fluconazole, and voriconazole in an in vitro model of *Candida* catheter-related bloodstream infection. *Antimicrob. Agents Chemother.* **46**, 3499–3505.

12. Garcia-Sanchez, S., Aubert, S., Iraqui, I., Janbon, G., Ghigo, J. M., and d'Enfert, C. (2004) *Candida albicans* biofilms: a developmental state associated with specific and stable gene expression patterns. *Eukaryot. Cell* **3**, 536–545.

13. Leunisse, C., van Weissenbruch, R., Busscher, H. J., van der Mei, H. C., and Albers, F. W. (1999) The artificial throat: a new method for standardization of in vitro experiments with tracheo-oesophageal voice prostheses. *Acta Otolaryngol.* **119**, 604–608.

14. Schuster, M., Lohscheller, J., Kummer, P., Hoppe, U., Eysholdt, U., and Rosanowski, F. (2003) Quality of life in laryngectomees after prosthetic voice restoration. *Folia Phoniatr. Logop.* **55**, 211–219.

15. Mahieu, H. F., van Saene, H. K., Rosingh, H. J., and Schutte, H. K. 1986. *Candida* vegetations on silicone voice prostheses. *Arch. Otolaryngol. Head Neck Surg.* **112**, 321–325.

16. Palmer, M. D., Johnson A. P., and Elliott, T. S. (1993) Microbial colonization of Blom-Singer prostheses in postlaryngectomy patients. *Laryngoscope* **103**, 910–914.

17. Neu, T. R., Verkerke, G. J., Herrmann, I. F., Schutte, H. K., van der Mei, H. C., and Busscher, H. J. (1994) Microflora on explanted silicone rubber voice prostheses: taxonomy, hydrophobicity and electrophoretic mobility. *J. Appl. Bacteriol.* **76**, 521–528.

18. Natarajan, B., Richardson, M. D., Irvine, B. W., and Thomas, M. (1994) The Provox voice prosthesis and *Candida albicans* growth: a preliminary report of clinical, mycological and scanning electron microscopic assessment. *J. Laryngol. Otol.* **108**, 666–668.

19. Van den Hoogen, F. J., Oudes, M. J., Hombergen, G., Nijdam, H. F., and Manni, J. J. (1996). The Groningen, Nijdam and Provox voice prostheses: a prospective clinical comparison based on 845 replacements. *Acta Otolaryngol.* **116**, 119–124.

20. Ackerstaff, A. H., Hilgers, F. J., Meeuwis, C. A., van der Velden, L. A., van den Hoogen, F. J., Marres, H. A., Vreeburg, G. C., and Manni, J. J. (1999) Multi-institutional assessment of the Provox 2 voice prosthesis. *Arch. Otolaryngol. Head Neck Surg.* **125**, 167–173.

21. Amann, R., Fuchs, B. M., and Behrens, S. (2001) The identification of microorganisms by fluorescence in situ hybridisation. *Curr. Opin. Biotechnol.* **12**, 231–236.

22. Aoi, Y. (2002) In situ identification of microorganisms in biofilm communities. *J. Biosci. Bioeng.* **94**, 552–556.

23. Thurnheer, T., Gmur, R., and Guggenheim, B. (2004) Multiplex FISH analysis of a six-species bacterial biofilm. *J. Microbiol. Methods* **56**, 37–47.

24. Buijssen, K. J., Harmsen, H. J., van der Mei, H. C., Busscher, H. J., and van der Laan, B. F. (2007) Lactobacilli: Important in biofilm formation on voice prostheses. Otolaryngol. Head Neck Surg. 137, 505–507.
25. Oosterhof, J. J., Buijssen, K. J., Busscher, H. J., van der Laan, B. F., and van der Mei, H. C. (2006) Effects of quaternary ammonium silane coatings on mixed fungal and bacterial biofilms on tracheoesophageal shunt prostheses. Appl. Environ. Microbiol. 72, 3673–3677.
26. Van der Mei, H. C., Free, R. H., Elving, G. J., van Weissenbruch, R., Albers, F. W., and Busscher, H. J. (2000) Effect of probiotic bacteria on prevalence of yeasts in oropharyngeal biofilms on silicone rubber voice prostheses in vitro. J. Med. Microbiol. 49, 713–718.
27. Elving, G. J., van der Mei, H. C., Busscher, H. J., van Weissenbruch, R., and Albers, F. (2003) Influence of different combinations of bacteria and yeasts in voice prosthesis biofilms on air flow resistance. Antonie Van Leeuwenhoek 83, 45–55.
28. Amann, R. I., Binder, B. J., Olson, R. J., Chisholm, S. W., Devereux, R., and Stahl D. A. (1990) Combination of 16S rRNA-targeted oligonucleotide probes with flow cytometry for analyzing mixed microbial populations. Appl. Environ. Microbiol. 56, 1919–1925.

Chapter 7

Conditions for Optimal *Candida* Biofilm Development in Microtiter Plates

Bastiaan P. Krom, Jesse B. Cohen, Gail McElhaney-Feser, Henk J. Busscher, Henny C. van der Mei, and Ronald L. Cihlar

Abstract

Development of *Candida* spp. biofilms on medical devices such as catheters and voice prosthesis has been recognized as an increasing clinical problem. Simple device removal is often impossible, while in addition, resulting candidal infections are difficult to resolve due to their increased resistance to many antifungal agents. Susceptibility studies of clinical isolates are generally performed according to the CLSI standard, which measures planktonic cell susceptibility, but similar standards have not been designed or applied to testing of cells growing within a biofilm. As consistent biofilms from many strains are more difficult to simultaneously obtain and analyze than are independent planktonic cultures, any standard assay must address these concerns. In the present chapter, optimized conditions that promote biofilm formation within individual wells of microtiter plates are described. In addition, the method has proven useful in preparing *C. albicans* biofilms for investigation by a variety of microscopic and molecular techniques.

Key words: Biofilm, microtiter plate assay, susceptibility testing, confocal laser scanning microscopy (CLSM), green fluorescent protein.

1. Introduction

Microbial biofilms are structured communities of cells surrounded by a self-produced matrix of extracellular polymeric substance (EPS). Biofilms usually develop at the interface of an aqueous medium and solid surface, and are of particular interest due to their contribution to microbial pathogenesis, as well as the clinical problems presented by the increased resistance of resident cells to antimicrobial agents. In order to determine the susceptibility of bacteria and fungi to antimicrobial drugs, standard assays have been catalogued by the Clinical and Laboratory Standards

Institute (CLSI; formerly National Committee for Clinical Laboratory Standards (NCCLS)), and in most cases, these have been developed exclusively for planktonically growing cells.

It is clear, however, that similar to bacteria, many *C. albicans* infections are due directly or indirectly to biofilm formation *(1, 2)*. As a consequence, results of currently used susceptibility assays for *C. albicans* may not be a true reflection of *in vivo* realities, and thus the use of suitable assays that measure the sensitivity of *C. albicans* cells growing in biofilms is desirable. This chapter describes conditions that promote optimal *C. albicans* biofilm formation in individual wells of microtiter plates *(3)*. Biofilms prepared can then be used in a high-throughput manner in susceptibility assays (sessile minimal inhibitory concentration (SMIC)) or other types of investigations that require concurrent development of independent biofilms. In the current chapter, methods to determine sensitivity of cells growing in biofilms to the antifungal agents amphotericin B (AmpB) and ketoconazole are described. In addition, the utility of the method in gene expression studies is also demonstrated.

2. Materials

2.1. Strains and Media

1. *C. albicans* SC5314 *(4)* and *C. albicans* HB12 *(5)*.
2. Yeast peptone agar plates with 2% glucose (YPD).
3. RPMI1640 (GIBCO BRL) buffered with 0.165 M 3-[*N*-morpholino]propanesulfonic acid (MOPS) (Sigma-Aldrich, USA) was prepared according to NCCLS 27-A2 standard *(6)*.
4. Difco™ yeast nitrogen base w/o amino acids (YNB) (Becton, Dickinson Co., USA) was prepared according to the manufacturers instructions and supplemented with 1% w/v glucose.
5. The pH of the medium was adjusted to 7 with 1 M KOH.
6. The medium was filter sterilized (*see* **Note 1**).

2.2. Biofilm Development

1. Phosphate-buffered saline (PBS) (10 mM potassium phosphate, 150 mM NaCl, pH 7.0).
2. Flat-bottom, low-evaporation tissue culture 96-well microtiter plate (Falcon™, Becton, Dickinson Co., USA).
3. Fetal calf serum (FCS) (Sigma-Aldrich, USA), alternatively, fetal bovine serum (Sigma-Aldrich, USA) can be used with identical results.

2.3. Biofilm Susceptibility Assay

1. AmpB and ketoconazole were obtained from Sigma-Aldrich (USA), and stock solutions were prepared in DMSO as described in NCCLS 27-A2.

2. MTT (3-(4,5-dimethyl-2-thiazolyl)-2,5-diphenyl-2H-tetrazolium bromide) is reduced by metabolically active cells to an insoluble dark purple crystal (Abs_{max} at 550 nm) (Sigma-Aldrich, USA).
3. MTT (0.5 mg/mL) was dissolved in PBS containing 1% glucose and 10 μM menadione (dissolved in acetone) (*see* **Note 2**).
4. Acid isopropanol (5% 1 M HCl in isopropanol).
5. Tecan platereader, Model GENios, or other appropriate reader (optional).

2.4. Confocal Laser Scanning Microscopy

1. Confocal laser scanning microscope model LEICA TCS SP2 (Leica Microsystems, Heidelberg GmbH, Heidelberg, Germany).
2. Lasers: Arg laser.
3. Wavelength: excitation 488 nm emission 514 (FITC settings).

3. Methods

Candida biofilm formation occurs in a well-orchestrated fashion, and is described by Chandra and coworkers *(7)*. Initial transient adhesion of yeast cells is followed by permanent adhesion. Adhesion involves physicochemical interactions and specific protein–protein interactions. In the method described here, the latter is achieved by preparing a conditioning layer of serum proteins on the surface of the wells. Adhesion is followed by germ-tube formation and further cellular differentiation into hyphae and pseudohyphae. During maturation, EPS is produced and the biofilm obtains its slimy appearance and more importantly, induces resistance to AmpB *(8)*.

The use of colorimetric assays, such as XTT(2,3-*bis*(2-methoxy-4-nitro-5-sulfophenyl)-5-[(phenylamine) carbonyl]-2H-tetrazolium hydroxide) *(9)*, or MTT, (3-(4,5-dimethyl-2-thiazolyl)-2,5-diphenyl-2H-tetrazolium bromide) to determine cell viability and thereby drug susceptibility of *Candida* cells in biofilms is widely accepted. MTT is reduced to insoluble purple formazan crystals, while XTT is reduced to water-soluble formazan. When using the MTT assay, it is required to solubilize the crystals prior to the measurement (*see* **Note 3**).

Comparison of MTT reduction as determined using the described method and plating have shown an excellent correlation between both MTT reduction and plating (**Fig. 7.1**).

3.1. Biofilm Growth

1. Overnight cultures of *C. albicans* grown in YPD are harvested by centrifugation (5 min 6000 × g).
2. Cells are washed with PBS and standardized to an optical density of 1×10^7 CFU/mL as determined by OD_{600} determination (*see* **Note 4**).
3. The wells are conditioned with 50% FCS in PBS for at least 30 min at room temperature.

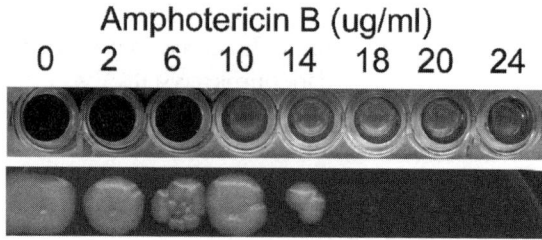

Fig. 7.1. Verification of the use of MTT reduction to assay viability. Prior to solubilization with acid isopropanol, 5 μL of biofilm (resuspended in residual liquid) was transferred to a YPD agar plate and incubated at 30°C for 24 h. In all our experiments, there was very good agreement between the MTT results and the plating.

4. The excess FCS is aspirated and wells are rinsed once with 200 μL PBS.
5. Cell suspension of 100 μL is added to each well.
6. The plate is incubated statically at 37°C for 90 min.
7. Nonadherent cells are removed by aspiration and wells are washed with PBS twice to remove loosely associated cells.
8. Biofilm growth is initiated by addition of 200 μL YNB to each well and subsequently incubated at 37°C for the desired amount of time (generally 24–48 h) (*see* **Note 5**).

3.2. Biofilm Susceptibility Assay

1. Biofilms are washed once with 200 μL PBS.
2. To each well containing a washed biofilm, 190 μL buffer or medium is added.
3. A 20-fold stock solution of AmpB or ketoconazole of 10 μL is added.
4. Biofilms are incubated at 37°C for an additional 24–48 h.

3.3. Detection of Biofilm Viability

1. After incubation, the biofilms are washed once with 200 μL PBS.
2. MTT of 100 μL is added to each well (*see* **Note 2**)
3. The plate is incubated at 37°C for 30 min.
4. The MTT reaction is terminated by aspiration of all solution.
5. The crystals are solubilized in 100 μL acid isopropanol. Efficient solubilization can be achieved by vigorous pipetting or shaking the plate for 10–15 min.
6. Killing efficacies can be monitored by eye or quantitatively by measuring the Abs at 550 nm (after transferring the solution to new 96-well plates).

In the standard susceptibility assay (the minimal inhibitory concentration (MIC) is determined), the inhibitor is included in the growth medium. To determine the MIC for biofilms, also referred to as the SMIC, biofilms have to develop in the absence of inhibitor. After washing, the biofilms are incubated in different buffers, or media, with inhibitor. This allows us to investigate the influence of different parameters on the killing process. Biofilms were grown for 48 h, washed with PBS, and incubated in different media and buffers with different pHs and buffer composition. The SMIC for AmpB was determined after 24-h incubation at 37°C.

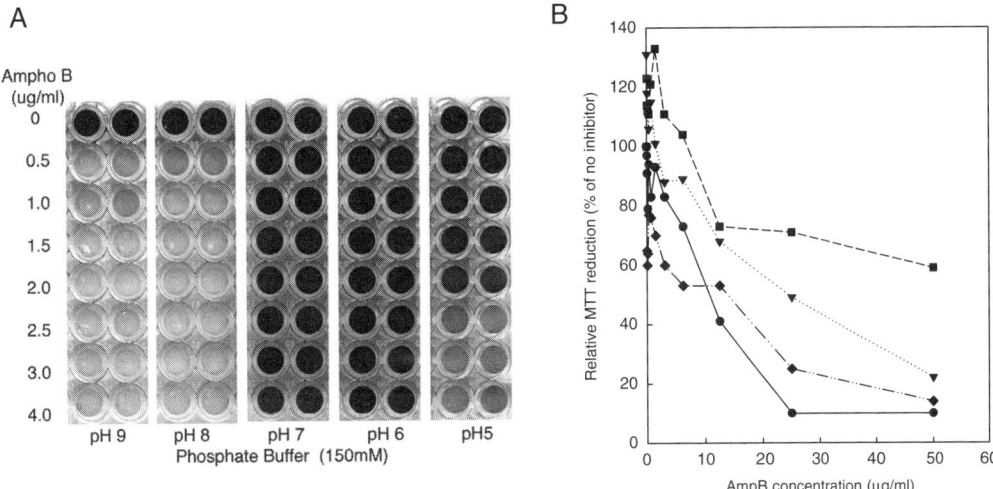

Fig. 7.2. **Panel A** shows the results of the MTT assay on *Candida* biofilms exposed to AmpB in 150 mM potassium phosphate buffer with different pH values. **Panel B** shows residual MTT reduction ability (percentage of no inhibitor) for biofilms exposed to increasing AmpB concentrations for 24 h at 37°C in YNB of different pH (◆ = pH 5, ■ = pH 6, ▼ = pH 7, and ● = pH 8).

No striking differences were observed when comparing YNB, MOPS, YNB MOPS, and RPMI MOPS all at pH 7 (not shown). Interestingly, the killing by AmpB appears to be very dependent on pH: lower pH yields lower SMIC values. Striking is that in 150 mM potassium phosphate buffer, at both pH 8 and pH 9, the SMIC is extremely low to below 1.0 µg/mL (**Fig. 7.2A**). In YNB, the pH effect is less dramatic, but still present (**Fig. 7.2B**). This increased sensitivity at higher and lower pHs was not observed when biofilms were incubated with ketoconazole (data not shown). Since ketoconazole is a cytostatic rather then a cytotoxic drug, we tested the efficacy of another cytotoxic drug, nystatin. Incubation of biofilms with increasing concentrations of nystatin at increasing pHs resulted in a pattern similar to the two azoles tested; there is no effect of pH on the SMIC value. It is, therefore, likely that the observed pH effect is specific for AmpB.

3.4. Confocal Laser Scanning Microscopy

1. Biofilms of *C. albicans* HB12 (or appropriate strain under study) are grown in 12-well plates as described above.
2. After 4 days incubations at 37°C, the biofilms are washed once with PBS to remove cells that have detached.
3. PBS is drained from the biofilms and the 12-well plate is either placed upside down under the microscope (using a long working distance lens) or is submerged in PBS in combination with a water lens (which can be inserted into overlying liquids).

This method of biofilm growth is also applicable to 24-, 12-, and 6-well culture plates for use with other techniques such as CLSM, or for mRNA isolation. **Figure 7.3** shows CLSM imaging of a GFP expressing *C. albicans* biofilm. GFP was fused to the

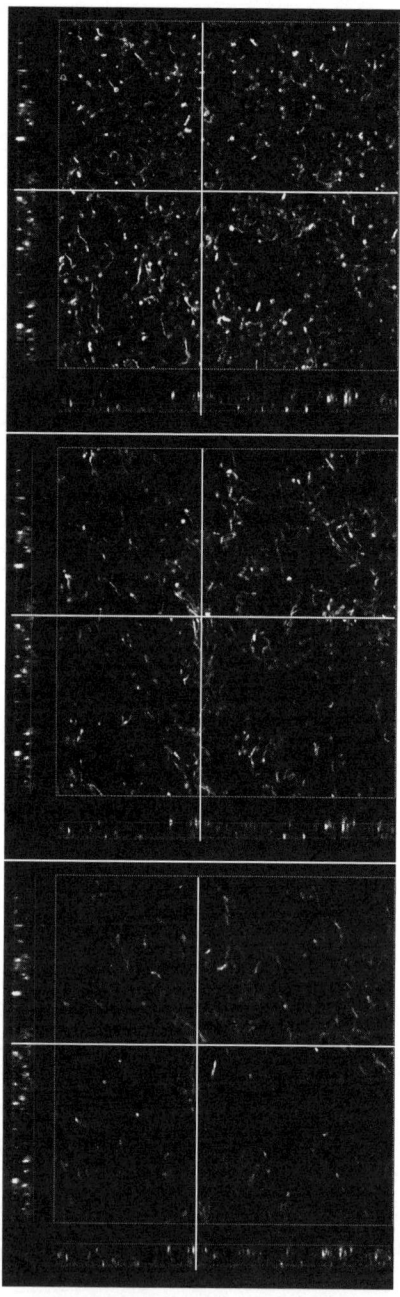

Fig. 7.3. CLSM images of GFP expressing *C. albicans* biofilms. Images were taken in cross-sections. The *three panels* show cross-sections at the indicated location of the biofilm going from the *bottom* (*top panel*) to the *top* (*bottom panel*) of the biofilm. It should be noted that only GFP expressing cells, i.e., those cells that express HWP1 (hyphal cells), are visible!

promoter of the hyphal specific gene HWP1 *(5)*. Biofilms were grown in 12-well plates as described and viewed using a LEICA TCS SP2 (Leica Microsystems, Heidelberg GmbH, Heidelberg, Germany) without further staining. Cells expressing HWP1 are clearly visible, in contrast to cells that do not express HWP1 (yeast cells) that are not visible in this image. From these images, it is clear that hyphal cells are prominent in the bottom layers of the biofilm, but are less abundant in the top layers.

4. Notes

1. Alternatively, the medium without glucose can be sterilized in an autoclave after which the sterile glucose can be added from a sterile 20% stock solution.
2. For optimal results, the PBS + glucose should be prewarmed to 37°C.
3. Kuhn and coworkers *(10)* have optimized the use of the XTT assay and showed that although XTT reduction results in a water-soluble formazan, it is recommended to solubilize the biofilm with DMSO prior to absorbance measurements. For both assays, it is important to notice that it is difficult to compare results from different isolates directly because the rate of MTT or XTT metabolism shows strain-to-strain variation. However, for susceptibility studies, this is not relevant since a drug-free control is included for each individual strain.
4. For *C. albicans* strain SC5314, $OD_{600}=1$ represents approximately 10^7 CFU/mL. Although the density at which adhesion is allowed matters to some extent, for susceptibility assays this is not as important. In our experience, this relation between OD_{600} and CFU is a good estimate for cell density of other clinical isolates.
5. Biofilm formation is influenced by agitation; especially EPS production is induced by agitation *(11)* and consequently elevated resistance might be observed due to penetration differences *(12)*. Therefore, the influence of static versus shaken incubation should be taken into account. No distinct effects of static versus shaken incubation was observed when the biofilms were grown in the 96-well plates, possibly because not enough medium flow is established in these small wells. However, using wells with larger surface area's shear, induced EPS formation have been obtained.

Acknowledgments

The authors thank Dr. Paula Sundstrom for the use of the *C. albicans* HB12 strain carrying the P_{HWP1}-GFP fusion.

References

1. Douglas, L. J. (2002) Medical importance of biofilms in *Candida* infections. *Rev. Iberoam. Micol.* **19**, 139–143.
2. Kojic, E. M., and Darouiche R. O. (2004) *Candida* infections of medical devices. *Clin. Microbiol. Rev.* **17**, 255–267.
3. Krom, B. P., Cohen, J. B., McElhaney Feser, G. E., and Cihlar, R. L. (2007) Optimized candidal biofilm microtiter assay. *J. Microbiol. Methods.* **68**(2), 421–423.
4. Fonzi, W. A., and Irwin, M. Y. (1993) Isogenic strain construction and gene mapping in *Candida albicans. Genetics* **134**, 717–728.
5. Staab, J. F., Bahn, Y. S., and Sundstrom, P. (2003) Integrative, multifunctional plasmids for hypha-specific or constitutive expression of green fluorescent protein in *Candida albicans. Microbiology* **149**, 2977–2986.
6. NCCLS. (2002) Reference Method for Broth Dilution Antifungal Susceptibility Testing of Yeasts: Approved Standard—Second Edition. NCCLS document M27-A2. NCCLS, Wayne, Pennsylvania, USA.
7. Chandra, J., Kuhn, D. M., Mukherjee, P. K., Hoyer, L. L., McCormick, T., and Ghannoum, M. A. (2001). Biofilm formation by the fungal pathogen *Candida albicans*: development, architecture, and drug resistance. *J. Bacteriol.* **183**, 5385–5394.
8. Al Fattani, M. A., and Douglas, L. J. (2006) Biofilm matrix of *Candida albicans* and *Candida tropicalis*: chemical composition and role in drug resistance. *J. Med. Microbiol.* **55**, 999–1008.
9. Hawser, S. P., Norris, H., Jessup, C. J., and Ghannoum, M. A. (1998). Comparison of a 2,3-bis(2-methoxy-4-nitro-5-sulfophenyl)-5-[(phenylamino)carbonyl]-2H-t etrazolium hydroxide (XTT) colorimetric method with the standardized National Committee for Clinical Laboratory Standards method of testing clinical yeast isolates for susceptibility to antifungal agents. *J. Clin. Microbiol.* **36**, 1450–1452.
10. Kuhn, D. M., Balkis, M., Chandra, J., Mukherjee, P. K., and Ghannoum, M. A. (2003). Uses and limitations of the XTT assay in studies of *Candida* growth and metabolism. *J. Clin. Microbiol.* **41**, 506–508.
11. Hawser, S. P., Baillie, G. S., and Douglas, L. J. (1998) Production of extracellular matrix by *Candida albicans* biofilms. *J. Med. Microbiol.* **47**, 253–256.
12. Baillie, G. S., and Douglas, L. J. (2000). Matrix polymers of *Candida* biofilms and their possible role in biofilm resistance to antifungal agents. *J. Antimicrob. Chemother.* **46**, 397–403.

Part III
Virulence Measurements: *In vitro*, *ex vivo*, and *in vivo*

Chapter 8

Animal Models of Candidiasis

Cornelius J. Clancy, Shaoji Cheng, and Minh Hong Nguyen

Abstract

Animal models are powerful tools to study the pathogenesis of diverse types of candidiasis. Murine models are particularly attractive because of cost, ease of handling, technical feasibility, and experience with their use. In this chapter, we describe methods for two of the most popular murine models of disease caused by *Candida albicans*. In an intravenously disseminated candidiasis (DC) model, immunocompetent mice are infected by lateral tail vein injections of a *C. albicans* suspension. Endpoints include mortality, tissue burdens of infection (most importantly in the kidneys, although spleens and livers are sometimes also assessed), and histopathology of infected organs. In a model of oral/esophageal candidiasis, mice are immunosuppressed with cortisone acetate and inoculated in the oral cavities using swabs saturated with a *C. albicans* suspension. Since mice do not die from oral candidiasis in this model, endpoints are tissue burden of infection and histopathology. The DC and oral/esophageal models are most commonly used for studies of *C. albicans* virulence, in which the disease-causing ability of a mutant strain is compared with an isogenic parent strain. Nevertheless, the basic techniques we describe are also applicable to models adapted to investigate other aspects of pathogenesis, such as spatiotemporal patterns of gene expression, specific aspects of host immune response and assessment of antifungal agents, immunomodulatory strategies, and vaccines.

Key words: Candida, *Candida albicans*, mouse, disseminated candidiasis, oral candidiasis, virulence, pathogenesis.

1. Introduction

Animal models are powerful tools for studying the pathogenesis of infectious diseases. By mimicking the complicated interactions between pathogens and human hosts at various tissue sites, such *in vivo* models offer advantages over less dynamic *in vitro* assays of pathogenesis. At the same time, animal models allow investigators to control specific pathogen and host factors in ways not possible in studies of humans or human tissues and cells. As the clinical importance of *Candida* species has increased over recent years, numerous animal models of candidiasis have been developed and validated. Depending on the experimental objectives, feasibility in a given lab, cost, and relevance to a particular type of human candidiasis, investigators can choose from models that differ in the species of animal, *Candida* species and specific strain, method of inoculation, tissue site of infection, immune status of the animal, time course, and endpoints to be assessed.

Clearly, a review of the range of animal models is beyond the scope of this chapter. Rather, we will present methods for two murine models of disease caused by *Candida albicans* strain SC5314: (i) intravenously disseminated candidiasis (DC) in immunocompetent mice; and (ii) oral/esophageal candidiasis in steroid-treated mice. We have chosen this approach for several reasons. First, *C. albicans* is the leading cause of systemic and mucosal candidiasis and the best-studied of the *Candida* species in animal models. *C. albicans* SC5314 is the most commonly used strain in pathogenesis studies and the genome sequence is available. Second, mice offer significant advantages over other animals in cost, ease of handling, technical feasibility, and track record of utility. The models we describe employ mice that are cheap and readily available from animal supply companies. Third, DC and oral candidiasis are the diseases that receive the most clinical and research attention. Finally, the models we describe are well-established and have been widely used.

The techniques we present in this chapter are general enough that they can be readily adapted to other murine models, including those using differing strains of mice or various immunosuppressive regimens. We highlight several common murine models of disseminated and mucosal candidiasis in **Table 8.1**, including the two described in detail below. The models in **Table 8.1** are used most commonly for studies of *C. albicans* virulence, in which the disease-causing ability of a mutant strain defective in a given factor of interest is compared with an isogenic parent strain. Nevertheless, many of the basic techniques are also applicable to models used to investigate other aspects of pathogenesis, such as spatiotemporal patterns of gene expression, specific aspects of the host immune response, and assessment of antifungal agents, immunomodulatory strategies, and vaccines.

Table 8.1
Common murine models of candidiasis

Types of candidiasis	Types of mice	Immunosuppression	Inoculum/ route of infection	Endpoints	References
Disseminated					
Immunocompetent, DC induced by IV infection.	Normal adult	None	Range: 1×10^5–1×10^6, IV	– Survival study – Tissue burden/histopathology of kidney, liver, and spleen	(1)
Neutropenic, DC induced by IV infection.	Normal adult	– Preinfection: Cyclophosphamide 100 mg/kg, day −4 – Postinfection: Cyclophosphamide 100 mg/kg days 1, 3, and 5.	1×10^4, IV	– Survival study – Tissue burden/ histopathology of kidney, liver, and spleen	(2)
Neutropenic, DC induced by gastrointestinal infection	6-Dday-old infant	– Preinfection: none – Postinfection: Cyclophosphamide (0.2 mg/g) and cortisone acetate (1.25 mg) days 11 and 14	2×10^8, oral-intragastric	– Tissue burden/ histopathology of stomach, kidney, liver, and spleen (day 20).	(3, 4)
Oral/esophageal	Normal adult	– Preinfection: Cortisone acetate (225 mg/kg) day −1 – Postinfection: Cortisone acetate days 1 and 3	$\sim 10^7$, sublingual	Tissue burden/histopath of tongue, submandibular tissue, and esophagus (days 5–7).	(5)
Vulvovaginal	CBA/J (H-2^k)	– Preinfection: Estradiol valerate 0.02 mg day −3 – Postinfection: Estradiol valerate 0.02 mg day 4	5×10^4, intravaginal	Vaginal lavages (days 4 and 7).	(6)

2. Materials

2.1. Murine Model of DC

2.1.1. Candida Strains

C. albicans SC5314 and other strains can be maintained on Sabouraud dextrose agar (SDA) plates at 4°C for up to 2 weeks. For long-term storage, log-phase *Candida* cells are saved in 25% glycerol in yeast peptone dextrose (YPD; 1% (w/v) yeast extract, 1%. (w/v) peptone and 2% (w/v) glucose) at –70°C (*see* **Note 1**).

2.1.2. Mice

Four- to six-week-old (approximately 20–25 g) adult immunocompetent mice (e.g., ICR, Balb/C mice; available from numerous animal supply companies).

2.1.3. Equipment (*see* **Note 2**)

1. Mouse tail illuminator (such as available from DAN-KAR Corp., MA, USA). This is optional and will be used only for the DC model.

2.1.4. Agar, Media, Reagents, Drugs, etc.

1. SDA (Difco/BRL, Kansas City, MO, USA).
2. YPD medium.
3. 85% Sterile saline.
4. Ampicillin, amikacin (Fisher Scientific). Ampicillin 1000 × stock (100 mg/mL) is prepared in 70% ethanol, and amikacin 1000 × stock (60 mg/mL) in distilled water. Stocks are stored at –20°C.

2.1.5. Supplies

1. Disposable 50 mL plastic centrifuge tubes.
2. Disposable 5 mL plastic tubes.
3. Syringe (0.5–1 mL) and needles (27–30 gauge).
4. Anatomy dissecting kit (available from multiple dealers; should contain dissecting scissors, scalpels, scalpel blades, forceps, and dissecting probes).
5. Disposable plastic pipettes (10, 200, and 1000 μL), inoculating loops (10 μL).
6. Flasks (250 mL, Pyrex).
7. Hemacytometer (Fisher Scientific).
8. Tissue homogenizer (Cat.7727-15, Pyrex).
9. Petri dishes (100 × 15 mm, Fisher Scientific).

2.2. Murine Model of Oral/Esophageal Candidiasis

Materials are as listed above for the model of DC, plus the followings.

2.2.1. Agar, Media, Reagents, Drugs, etc.

1. Cortisone acetate (Sigma-Aldrich, St Louis, MO, USA). Prepare fresh 40 mg/mL suspension with saline containing 0.1% Tween 80.
2. Tetracycline hydrochloride (Fisher Scientific). Prepare fresh 0.5 mg/mL suspension with distilled water. Store in the dark.
3. Pentobarbital sodium (50 mg/mL, Abbott Laboratories, North Chicago, IL, USA). Store at room temperature. Dilute to 20 mg/mL with saline before use.

2.2.2. Supplies

1. Calcium alginate urethral swabs (type 4 Calgiswab; Puritan Medical Products Company LLC, Guilford, ME) (*see* **Note 3**).
2. Syringe (1 mL) and needles (25 gauge).

3. Methods

3.1. Murine Model of DC

3.1.1. Preparation of Inoculum

1. Using a sterilized loop, streak the *C. albicans* frozen stock to isolation on an SDA plate (*see* **Note 4**). Incubate at 30°C.
2. Two days prior to the inoculum preparation, pick a single colony and streak to isolation on an SDA plate. Incubate at 30°C. Make sure the plate shows homogenous growth and no contamination (*see* **Note 5**) (**Fig. 8.1**).
3. One day prior to the inoculum preparation, pick a single colony from the SDA plate using the inoculum loop and inoculate into a 250-mL flask containing 50 mL of YPD.
4. Incubate at 30°C with shaking at 150 rpm for at least 18 h (*see* **Note 6**).
5. Examine an aliquot of the culture under a microscope for *Candida* cell morphology and lack of bacterial contamination; >95% of *Candida* cells should be blastoconidia (*see* **Note 7**) (**Fig. 8.2**).

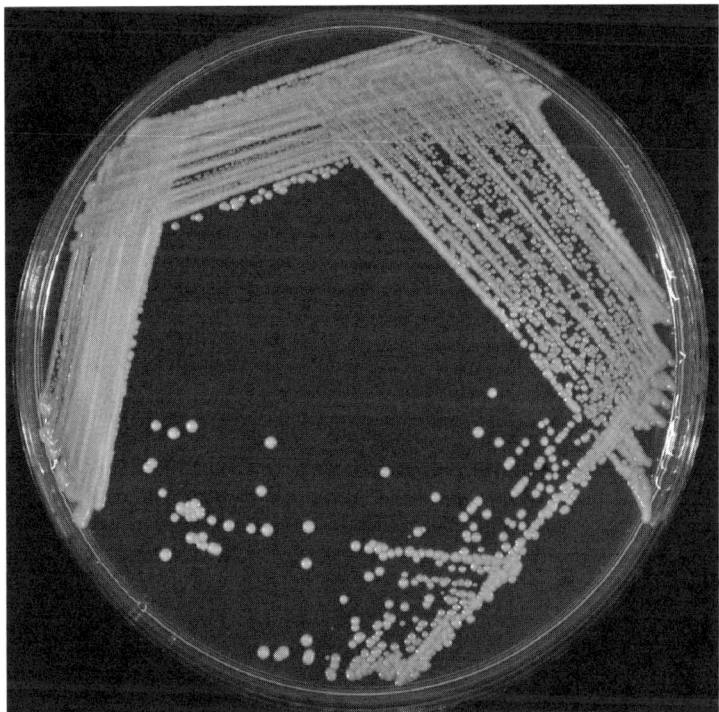

Fig. 8.1. *C. albicans* colonies on SDA (*see* **Note 5**).

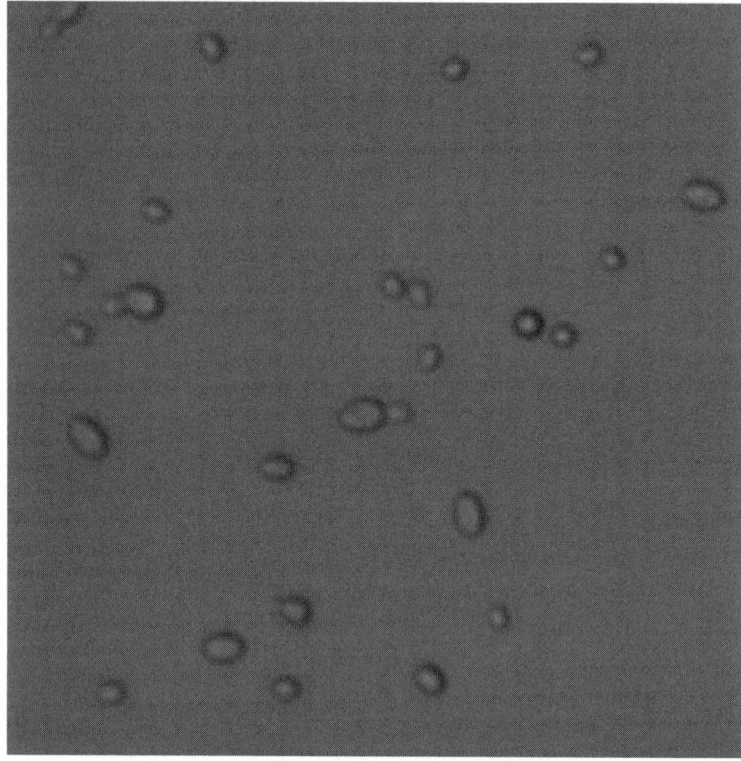

Fig. 8.2. Morphology of *C. albicans* cells following inoculum preparation (*see* **Note 7**).

3.1.2. Washing Cells

1. Transfer the overnight culture into a 50-mL plastic disposable centrifuge tube, and centrifuge it at $1000 \times g$ for 20 min.
2. Discard the supernatant. Scoop five loopfuls of the pellet into a 5-mL culture tube containing 2 mL of 85% sterile saline solution.
3. Vortex the 2 mL cell suspension and centrifuge it at $1000 \times g$ for 10 min.
4. Discard the supernatant and wash the pellet with 2 mL 85% sterile saline solution at least three times. Vortex and centrifuge as described above.
5. After the third wash, dispense the pellet with 2 mL 85% sterile saline solution and vortex.

3.1.3. Counting Cells (*see* **Note 8**)

1. Make three 1:10 serial dilutions in sterile saline (10^{-1}, 10^{-2}, 10^{-3}) from the 2 mL cell suspension using 15 mL culture tubes. The final volume in each tube is 5 mL.
2. Place a small drop of cell suspension from the 10^{-3} dilution tube at the edge of the coverslip on the hemocytometer.
3. Count the number of cells in the central part of the grid, and calculate the cell density per milliliter of volume.

3.1.4. Making Inoculum

1. Calculate the desired concentration and total volume of the inoculum.
   For example, *Goal*: 1×10^6 CFU per 0.2 mL of inoculum (for each mouse).
   The concentration (CFU/mL) of the inoculum is, therefore, 1×10^6 (CFU/mouse)/0.2 (mL/mouse) = 5×10^6 (CFU/mL). If 20 mice are to be included in the study, the total volume (mL) is 0.2 (mL/mouse) × 20 (mice) = 4 mL.
   In this event, make it 6 mL to account for possible volume loss.

2. Calculate the amount of cell suspension in the 10^{-3} tube needed to make 6 mL of 5×10^6 CFU/mL inoculum.
   For example, if the cell concentration of the 10^{-3} tube is 7.58×10^6 CFU/mL, 7.58×10^6 (CFU/mL) × X (mL) = 5×10^6 (CFU/mL) × 6 (mL) ∴ X = 3.96 mL.
   Therefore, 3.96 mL of the cell suspension in the 10^{-3} tube is needed to make a total of 6 mL of (5×10^6 CFU/mL) inoculum.

3. Add 3.96 mL of the suspension in the 10^{-3} tube to a tube containing 2.04 mL of sterile saline to make 6 mL of 5×10^6 CFU/mL inoculum.

3.1.5. Checking the Inoculum

1. Make 1:10 serial dilutions in sterile saline (10^{-3}, 10^{-4}) from the 5×10^6 CFU/mL inoculum. The final volume in each tube is 1 mL.
2. Spread 0.1 mL of each dilution onto an SDA plate and incubate at 30°C.
3. Count the number of colonies growing after 24 h of incubation. Acceptable error is 5–10%.

3.1.6. Tail Vein Injection

1. Restrain the mouse and warm the tail with mouse-tail illuminator (*see* **Note 9**).
2. Rotate the tail slightly to visualize the lateral tail vein. Disinfect the injection site with isopropyl alcohol and insert the needle (27–30 gauge) bevel-up into the vein at a slight angle.
3. Inject 200 μL of the inoculum slowly and watch for clearing of the lumen of the vein. The plunger should meet no resistance (*see* **Note 10**).
4. Upon completion, remove the needle and apply pressure to the injection site.

3.1.7. Following Mice

1. Check the mice three times a day for any signs of distress: decreased food and water consumption, weight loss, self-imposed isolation/hiding, rapid breathing, opened-mouth

breathing, increased/decreased movement, abnormal posture/positioning, dehydration, twitching, trembling, and tremor.
2. Mice showing signs of distress accompanied by any of the following are defined as moribund: impaired ambulation (unable to reach food and water), evidence of muscle atrophy or signs of emaciation, lethargy (drowsiness, aversion to activity, lack of physical or mental alertness), prolonged anorexia, difficulty breathing, and neurological disturbances.
3. For survival studies, euthanize the mice when they look moribund by CO_2 asphyxiation followed by cervical dislocation. The date of death is recorded as the following day (*see* **Note 11**). Assess the differences in survival between the groups of mice using Log-Rank test; p-value < 0.05 is considered significant (*see* **Note 12**).
4. For tissue burden and histopathology studies, the mice will be euthanized at a predetermined time point (*see* **Note 13**).

3.1.8. Dissection

1. Pin the euthanized mouse down with the belly facing up and wet the fur of the chest and abdomen with ethanol.
2. Elevate the skin with sterilized forceps and cut along the ventral midline from the groin to the sternum. Next, make an incision from the start of the first incision downward to the knee on both sides of the mouse and pull the skin back on the sides.
2. Cut through the peritoneal muscle wall and open up the body cavity. The kidneys are located bilaterally on the dorsal wall of the abdominal cavity. The spleen is located behind the stomach.
3. Lifting up the sternum with forceps, puncture the diaphragm and cut through each side of the sternum up through the cervical girdle. Cut the diaphragm out in order to remove the liver (immediately below diaphragm in right upper quadrant of abdomen).
4. Small samples of desired organs are dissected for histopathology and stored immediately in the appropriate fixative. The remainder of the organ is weighed. After weighing, the organs are homogenized in 2 mL of ice-cold normal saline.
5. Serially dilute the homogenate in normal saline (1:10 dilutions). Plate 100 µL of each dilution onto SDA plates containing ampicillin (100 µg/mL) and amikacin (60 µg/mL). Incubate at 30°C for 24–48 h (*see* **Note 14**).
6. Count the number of colonies and calculate as CFU/g tissue. Assess the differences in tissue burden between two strains using the Wilcoxon Rank Sum test (p-value <0.05 is considered significant).

3.2. Murine Model of Oral/Esophageal Candidiasis

1. Prepare 1×10^8 CFU/mL inoculum, using the methods described above.
2. Add tetracycline hydrochloride to drinking water at a concentration of 0.5 mg/mL starting the day before infection. Allow mice to drink *ad lib*.
3. Give 225 mg/kg of cortisone acetate subcutaneously on the day before infection and days 1 and 3 postinfection (*see* **Note 15**).
4. Prior to infection, anesthesize the mice by intraperitoneal injection of pentobarbital sodium solution (*see* **Notes 16 and 17**).
5. Place a calcium alginate urethral swab saturated with a suspension containing 1×10^8 CFU/mL sublingually in the oral cavity for 2 h (*see* **Note 18**).
6. Remove the swab and continue to observe until the mice awaken and resume normal activity.
7. Mice are observed three times a day following infection for signs of distress, as described above.
8. Mice are generally followed for 5–7 days prior to euthanizing by CO_2 asphyxiation followed by cervical dislocation.
9. Dissect the esophagus and the mandibular soft tissue, including the tongue, free of teeth and bone.
10. Save a small sample for histopathology and weigh the rest of the tissue. Homogenize the tissue in 2 mL of ice-cold normal saline.
11. Serially dilute the homogenate in saline. Plate 100 μL of each dilution onto SDA containing ampicillin (100 μg/mL) and amikacin (60 μg/mL). Incubate at 30°C for 48 h.
12. Count colonies and calculate as CFU/g tissue (*see* **Note 19**). Assess the differences in tissue burdens between the two strains using the Wilcoxon Rank Sum test (p-value <0.05 is considered significant) (*see* **Note 20**).

4. Notes

1. Virulence studies typically compare isogenic *C. albicans* strains for one (or more) of three endpoints in mice with DC: mortality, tissue burden of infection, and histopathology of infected organs. Although mortality studies are most common, it is important to recognize that death is a complex and relatively crude endpoint that reflects multiple factors involving both the organism and the host *(7)*. For this reason, the best measurement of a strain's virulence potential also includes a careful assessment of disease at specific organs. In all virulence studies, the null mutant should be compared to a suitable parental strain and a gene reinsertion

strain. For the methods we describe, *C. albicans* SC5314 and CAI-12 (a derivative of SC5314 in which a copy of *URA3* has been reinserted in a *ura3* background) yield comparable results in mice.

2. These experiments require standard laboratory equipment, such as centrifuges, light microscope, analytical balance, shaking and nonshaking 30°C incubator, and vortexes.
3. These will be used to inoculate the oral cavities. They are superior to cotton swabs, which do not stay as tight when moistened with the inoculum preparation.
4. These are standard methods, as used for isolation of bacterial colonies.
5. *C. albicans* colonies are significantly larger than those of bacteria. On SDA, *C. albicans* is white to cream colored, smooth, glabrous, and yeast-like in appearance (*see* also **Fig. 8.1**).
6. Loosen the lid to allow air in, which facilitates growth.
7. Microscopic morphology should show spherical, budding yeast-like cells or blastoconidia, generally about 4 × 6 μm in size. Filamentous forms (pseudohyphae and hyphae) of strain SC5314 are not suitable for inoculation. They will be more prominent at higher temperatures, which is why inoculum preparation is performed at 30°C (*see* also **Fig. 8.2**).
8. *C. albicans* cells clump very easily. Vortex vigorously for at least 15 s before each aspiration.
9. The mouse-tail illuminator is handy if the investigator is less experienced with the tail vein inoculation procedure. The mouse is placed in a chamber and immobilized, which makes it easier to keep the tail still for inoculation. We find that the light and heat provided by the illuminator are generally less useful, since the tail veins are relatively large and easily visualized if prior attempts at inoculation have not been made. With experience, it is more rapid to forgo the illuminator. In this event, grab the mouse by the tail and suspend him over the edge of the cage so that his back is against the inner cage wall. Replace the cage lid over the tail, pulling the tail taut toward you (but not so taut that the color is drained from the tail and veins). While immobilizing the tail, perform inoculation with the other hand.
10. It is best to come at the vein slightly from the side to minimize rolling of the vein. Incorrect positioning of the needle will result in a slight bulge in the tail. If this occurs, remove the needle and repeat the process proximal to previous site. The key to knowing you are successfully inoculating rather than infiltrating the tail is the lack of resistance. The plunger should meet no resistance whatsoever, and you should see an immediate clearing of the vessel lumen proximal to the injection site. If there is any resistance or the tail begins to turn white, stop immediately and reattempt at a proximal site.

11. Mice infected with 1×10^6 CFU of strain SC5314 will generally begin to die on day 3, with 100% mortality within the first week. This is usually the standard inocula for comparative mortality studies. Inocula of 1×10^5 CFU of strain SC5314 typically delay the onset of death for several days and cause 100% mortality within 2 weeks.
12. Mortality experiments generally include 10 mice per group, which is a sufficient number to achieve statistical significance. In tissue burden experiments using the DC model, we usually include 12 mice per group.
13. It is not valid to compare tissue burdens at the time of death for mice dying on different days from DC. Since inocula of 1×10^6 CFU of strain SC5314 cause rapid and reliable death over several days, tissue burden and histopathology experiments are generally performed with lower concentrations.
14. The kidney is the principal target organ during DC. Candidal colony counts in the kidneys progressively increase during fatal infection and decrease during nonfatal infection. Generally, 10^6–10^8 *C. albicans* CFU/g of tissue can be obtained from the kidneys of moribund mice infected with strain SC5314. The spleen and liver are less commonly studied than the kidney; colony counts typically reach up to 10^4 CFU/g.
15. Cortisone acetate is insoluble in saline and water. Make a fresh 40 mg/mL suspension with saline containing 0.1% Tween 80. Mix well by rigorous vortexing and use 0.1 mL for each subcutaneous injection. Grasp the loose skin on the back of the mouse from ears to the legs and restrain the legs with your ring and little fingers. Disinfect the injection site and insert the needle (25 gauge) into the subcutaneous tissue in the lateral side of the abdominal wall. Aspirate prior to making the injection, and inject slowly upward toward the armpit of the mouse. Proper placement should yield no aspirate. The injection should be seen through the skin as a lump. In the event that you find mice difficult to restrain, briefly spinning the animal by the tail will make it dizzy and easier to hold still.
16. Dilute the pentobarbital sodium solution to 20 mg/mL with sterile water. Use 130–150 mg/kg for IP injection. In our experience, 140 mg/kg keeps mice of this age and weight asleep for approximately 2.5 h. We have very few deaths due to overdose with this method. Nevertheless, the mice should be observed closely during the entire period of anesthesia. In the event that a mouse stops breathing, light compression of the chest with a single finger is generally sufficient to resuscitate. We use pentobarbital because it reliably sedates the mice for the duration of the inoculation period. As an alternative to pentobarbital, injection IP with xylazine (8 mg/kg) and ketamine (11 mg/kg) (Phoenix Pharmaceuticals, St Joseph,

MO, USA) can be used. Using these agents, we often shorten the inoculation period to 75 min.

17. Grasp and restrain the mouse using methods similar to those used for the subcutaneous injection of cortisone acetate. Tilt the mouse so that the head is facing downward and its abdomen is exposed. Disinfect the injection site and insert the needle cranially into the abdomen at a 30–45° angle caudal to the umbilicus and lateral to the midline (i.e., in the lower quadrants of the abdomen). If nothing is aspirated, inject.

18. Prepare the inoculum solution and insert the swabs into the solution until saturated. This method results in an inoculum of approximately 100 µL (i.e., 1×10^7 CFU/mouse). Shake loose any solution by flicking wrist prior to inserting the swab into the oral cavity. Investigators have used inocula of 1×10^7 CFU/mL with this model as well.

19. Due to the range of data that might be obtained with this model, results are generally best if interquartile data are used. For this reason, we usually include ≥ 16 mice/group. In mice infected orally with *C. albicans* SC5314, 10^4–10^5 CFU/g of tissue is typical.

20. Mice will rarely die as a result of oral candidiasis in this model; such animals usually have visible oral thrush and emaciation. Mortality rates should be $\leq 5\%$. Tissue burdens and histopathology are the only endpoints usually assessed.

References

1. Cheng, S., Clancy, C. J., Checkley, M. A., Wozniak, K. L., Seshan, K. R., Jia, H. Y., Fidel, P., Cole, G., and Nguyen, M. H.. (2005) The role of *Candida albicans* NOT5 in virulence depends upon diverse host factors in vivo. *Infect. Immun.* 73, 7190–7197.
2. Netea, M. G., van Tits, L. J., Curfs, J. H., Amiot, F., Meis, J. F., van der Meer, J. W., and Kullberg, B. J. (1999) Increased susceptibility of TNF-alpha lymphotoxin-alpha double knockout mice to systemic candidiasis through impaired recruitment of neutrophils and phagocytosis of *Candida albicans*. *J. Immunol.* 163, 498–1505.
3. Cole, G. T., Seshan, K. R., Pope, L. M., and Yancey, R. J. (1988) Morphological aspects of gastrointestinal tract invasion by *Candida albicans* in the infant mouse. *J. Med. Vet. Mycol.* 26, 173–185.
4. Cole, G. T., Lynn, K. T., Seshan, K. R., and Pope, L. M. (1989) Gastrointestinal and systemic candidosis in immunocompromised mice. *J. Med. Vet. Mycol.* 2, 363–380.
5. Kamai, Y., Kubota, M., Kamai, Y., Hosokawa, T., Fukuoka, T., and Filler, S. G. (2002) Contribution of *Candida albicans* ALS1 to the pathogenesis of experimental oropharyngeal candidiasis. *Infect. Immun.* 70, 5256–5258.
6. Wozniak, K. L., Wormley, F. L., Jr., and Fidel, P. L., Jr. (2002) *Candida*-specific antibodies during experimental vaginal candidiasis in mice. *Infect. Immun.* 70, 5790–5799.
7. Casadevall, A. and Pirofski, L. (2001) Host-pathogen interactions: the attributes of virulence. *J. Infect. Dis.* 184, 337–344.

Chapter 9

Candida albicans Gene Expression in an *In Vivo* Infection Model

Michael D. Kruppa

Abstract

A general procedure is described for the analysis of gene expression of *Candida albicans* cultured in a mouse infection model. This technique involves first infecting mice with *Candida* and subsequently harvesting blood and other tissue at specific time points during infection. The tissues are homogenized and the infecting *Candida* isolated. Finally, RNA is extracted from recovered *Candida* cells and subjected to microarray analysis.

Key words: *Candida*, microarray, animal infection model, gene expression.

1. Introduction

In recent years, genome sequencing and development of microarray technology has allowed investigators to examine the transcript profiles of *Candida albicans* cells grown under a number of *in vitro* conditions *(1–7)*. Comparison of mutant strains to their cognate wild type has helped in elucidating transcriptional networks in *C. albicans* under *in vitro* conditions. In addition, many genes responsible for virulence/survival within the host have been identified based upon the inability of a mutant strain to cause death in an *in vivo* animal model system. Other genes which may contribute to virulence may be overlooked if they play only a minor role and, therefore, these may function in tandem with a network of other genes. As a surrogate approach, studies have been done using microarray technology with RNA recovered from *Candida* cells cultured in the presence of macrophage and neutrophils or their cell lines *(8–11)*. These *in vitro* studies have shown differences in the response of *C. albicans* to specific immune cell challenge, but are the results complementary to what occurs *in vivo*. In one approach to this question, studies have described the fractionation of human blood, followed by introduction of *C. albicans* and the demonstration that *Candida* transcriptional expression varies as a function of the different blood components challenged, e.g., neutrophils, macrophages, or plasma *(4, 8–10)*. Furthermore, it has been recently shown that mice infected with *C. albicans* can be used as a model to measure transcriptional changes in *Candida* isolated from either tissue or blood *(4, 12)*. In this context, the aim of this chapter is to describe a general approach to examine *C. albicans* gene expression in an *in vivo* model.

2. Materials

2.1. C. Albicans Culture Conditions for In Vivo and In Vitro Inoculation

1. Plate culture of *C. albicans* strains to be tested (*see* **Note 1**).
2. Yeast-extract–peptone–dextrose medium (YPD) for overnight culture of organisms at 37°C.
3. Wash/suspension buffer: 0.85% NaCl.
4. 6 × 100 mL YPD broth in 500 mL flasks.

2.2. In Vivo Culture

1. *Candida* overnight cultures grown at 37°C in YPD.
2. For organ or blood extraction: 40 mice, 24–27 g; use a mouse strain of choice such as Balb-C or Black Six mice (*see* **Note 2**).

2.3. Organ Processing

1. Mechanical tissue grinder.
2. Organs extracted from 10 infected mice (kidneys, spleen, etc.).
3. Sterile distilled water.
4. Filter discs (22 µm pores, 25 mm diameter) for removal of tissue debris.
5. RNase-free DNase.
6. Triton X-100.
7. Liquid nitrogen for immediate freezing of sample.
8. YPD plates.

2.4. RNase Treatment to Remove Contaminating Host RNA

1. RNase A (500 U/mL) and RNase T1 (20,000 U/mL) cocktail (Ambion Technologies) or acquired separately from other source.
2. SUPERase RNase inhibitor (20 U/µL) (Ambion).

2.5. Extraction of Candida RNA

1. Acid–Phenol.
2. Chloroform.
3. RNase-free water.
4. TES: 10 mM Tris-Cl pH7.5, 10 mM EDTA, 0.5% SDS.
5. Glass beads.
6. 3 M Sodium acetate, pH 5.2.
7. 100% Ethanol – ice cold.
8. Optional: RNase-free DNase.

2.6. RNA Amplification and Labeling

1. Total RNA extracted from *in vivo* and *in vitro* grown cells.
2. cDNA synthesis/ hybridization kit (sources: 3DNA technologies, Stratagene, GE Biosciences).
3. Optional: Cy3 dye for labeling of *in vitro* grown RNA sample (GE Biosciences) if using kit other than 3DNA.
4. Optional: Cy5 dye for labeling of *in vivo* grown RNA sample (GE Biosciences) if using kit other than 3DNA.
5. Other reagents as required from manufacture's notes.

2.7. Hybridization of Microarrays

1. Microarray chips purchased from a suitable source (Operon Technologies, Affymetrix, Biotechnology Research Institute, National Research Council, Montreal).
2. Prehybridization/hybridization buffer: 5X standard saline citrate (SSC), 0.1% SDS, 50X Denhardt's solution.
3. tRNA.
4. Denatured genomic DNA.
5. Wash buffer 1: 1X SSC, 0.2% SDS.
6. Wash buffer 2: 0.1X SSC, 0.2% SDS.
7. Wash buffer 3: 0.1X SSC.
8. Hybridization oven/water bath set at 42°C.

2.8. Gene Expression and Data Analysis

1. Genepix array scanner or other scanning system of choice.
2. Microarray software analysis package such as the TM4 package (TIGR), Genespring or other comparable software.

3. Methods

3.1. Preparation of In Vivo Cultures

1. Cultures of *Candida* are grown in YPD at 37°C (50 mL cultures in 250 mL Erlenmeyer flasks) overnight for 16–24 h. Cells are harvested and washed three times with 0.85% NaCl.

2. For organ extraction, cell densities are adjusted to 1×10^8 cells/mL in 0.85% NaCl. 1×10^6 Cells are injected into the tail vein of 10 neutropenic mice (for organ harvest) for each sample point. For blood harvest, 1×10^5 cells are injected into the tail vein of 10 non-neutropenic mice for each sample point (*see* **Note 3**).

3.2. Sample Harvest and Preparation

3.2.1. Organ Harvest

1. In general, mice are sacrificed at 5, 10, 15, and 30 h post-inoculation for organ extraction. Organs are removed from mice at each time point. Ten mice are necessary for each time point.
2. Organs are placed in 1 mL of water and homogenized using a mechanical grinder.
3. Homogenates are filtered through three to four plies of 22–25 μm pore discs to remove tissue debris. A fraction of the flow through should be diluted and plated onto YPD plates to determine the colony forming units (CFU) of each sample.
4. Flow through should be immediately frozen using liquid nitrogen.
5. The flow through is then thawed and 4 U/mL RNase-free DNase is added to it. Triton X-100 is also added to a final concentration of 1%. The mixture is then incubated at 37°C for 20 min.
6. The tissue suspension is then centrifuged at 500 × *g* for 5 min to remove any remaining tissue. Save the supernatant and discard the pellet.
7. A second centrifugation step at 4000 × *g* for 3 min is necessary to recover the yeast cells, which will be present in the pellet.
8. Suspend the pellet in 270 μL RNase-free water.
9. Mammalian RNA is removed from the pellet by adding 30 μL of 10X RNase buffer and 1 μL of a cocktail of RNase A and RNaseT1 to the sample. Incubate for 20 min at 37°C.
10. Inactivate the RNase by adding 20 μL of SUPERase RNase inhibitor (20 U/μL). Incubate for 20 min at 37°C.
11. Centrifuge the pellet at 4000 × *g* for 3 min and remove the supernatant. At this point, the *Candida* cell pellet may be stored at –70°C, or proceed with the RNA extraction.

3.2.2. Blood Harvest

1. Animals are bled 30 min, 1 h, and 2 h after tail vein infection. Mice are bled by intracardiac puncture for each time point, which should yield up to 0.5 mL of blood/mouse.

2. Blood cells are removed by hypotonic lysis. Sterile water of 10 mL is added to the sample, which will lyse the majority of red blood cells. The sample should then be vigorously agitated with the tissue grinder to further disrupt/lyse remaining cells.

3. Samples are then centrifuged for 5 min at 500 × g to pellet cell debris. Take a fraction of the supernatant, dilute, and plate onto YPD to determine the CFU of the sample.

4. Centrifuge the remaining supernatant for 5 min at 4000 × g to pellet the *Candida* cells.

5. Proceed to **Step 8** in **Section 3.2.1**.

3.2.3. *In Vitro* Culture Cell Harvest

1. For *in vitro* cultures, cells should be inoculated to an optical density$_{600}$ (OD$_{600}$) of 0.05 in 6 × 100 mL cultures in YPD and incubated with shaking at 250 rpm at 37°C. Cells are grown for up to 12 h and harvested every 2 h. A fraction of each sample should be diluted and plated to estimate the number of CFU on YPD. Each sample is then frozen and stored at −70°C until ready to extract the RNA.

3.3. RNA Extraction from Candida

1. *Candida* RNA is extracted with hot phenol *(13)*. The cell pellet is suspended in 300 μL TES.

2. Glass beads of 0.2 g is added to the tube.

3. Acid phenol of 300 μL is added to the tube.

4. Tubes are vortexed vigorously for 10 s and placed in a 60°C water bath for up to 1 h. Tubes should be vortexed every 10–15 min.

5. Samples are placed on ice for 5 min and then centrifuged for 5 min.

6. The top aqueous phase is transferred to another tube and a second phenol extraction is performed. Cells are vortexed for 10 s to mix and then placed on ice for 5 min. Tubes are then centrifuged for 5 min.

7. The aqueous phase from **Step 6** is transferred to another tube and 400 μL of chloroform is added. The sample is then vortexed and placed on ice for 5 min.

8. Centrifuge 5 min and transfer aqueous phase to a fresh tube (*see* **Note 4**). Sodium acetate of 30 μL, pH 5.2 is added and 3 volumes of ice-cold 100% ethanol is added. Vortex tube and place on ice for 5 min. Centrifuge for 5 min.

9. Discard supernatant and allow pellet to dry. Suspend RNA pellet in 50 μL RNase-free water and determine the concentration.

3.4. cDNA Synthesis, Sample Labeling, and Hybridization

1. The preferred method for labeling of samples is through the use of the 3DNA array kits due to ease of use and low background. Detection of RNA amounts as low as 5 μg total RNA can easily be obtained (see **Note 5**).

2. Set up a labeling reaction for each sample using anchor primers for cDNA synthesis as described in the 3DNA protocol. After cDNA synthesis is finished, and the RNA used for template degraded, cDNAs can be concentrated and combined with hybridization buffer, denatured and allowed to hybridize overnight at 42°C.

3. Arrays are washed according to the protocol with prewarmed wash buffers followed by a fixing step, which allows for the cDNAs to remain adhered to their target genes. The array is then air-dried and prepared for a second hybridization.

4. Second hybridization involves the hybridization of the two 3DNA dendrimer molecules, prelabeled with cy3 or cy5. These dendrimers contain complimentary sequences, which bind to the anchor primers that were used for cDNA synthesis.

5. A hybridization mixture containing the 3DNA labeled dendrimers is incubated for 10 min at 80°C, placed on the array containing the hybridized cDNAs, and finally incubated at 42°C for up to 3 h.

6. Microarrays are then washed according to the protocol described immediately above, dried, and then exposed to an array scanner such as Genepix 4000.

3.5. Data Analysis

1. After arrays have been scanned, the individual spots must be analyzed by grid analysis software, such as TIGR Spotfinder or another suitable product. This allows for quantification of data and determination of quality of hybridization.

2. After spots have been quantified, they must be standardized by statistical analysis, generally a LOWESS algorithm. This is followed by standard deviation regularization of the array data. Another important aspect of data analysis is comparative analysis by flip-dye labeling.

3. After data standardization, individual arrays can be compared using an analysis program such as TIGR MEV, which allows sorting of the individual genes in a hierarchical order, clustering according to expression level and according to function, if the latter has been previously defined.

4. Notes

1. *C. albicans* cultures should be from fresh culture plates inoculated from freezer stocks.

2. Mice should be rendered neutropenic by administration of cyclophosphamide (150 mg/kg for 4 days or 100 mg/kg for 1 day). A genetically neutropenic mouse strain may also be used.

3. For each sample point, 10 mice are generally needed; however, this must be scaled up or down depending upon the hybridization detection system used.

4. Samples may be treated with RNase-free DNase for 20 min to reduce any genomic DNA contamination from both tissue and *C. albicans*

5. Other kits are available from Genisphere 3DNA that allows levels of detection to as low as 25–100 ng total RNA.

References

1. Bensen, E. S., Martin, S. J., Li, M., Berman, J., and Davis, D. A. (2004) Transcriptional profiling in *Candida albicans* reveals new adaptive responses to extracellular pH and functions for Rim101p. *Mol. Microbiol.* **54**, 1335–1351.

2. Kadosh, D. and Johnson, A. D. (2005) Induction of the *Candida albicans* filamentous growth programs by relief of transcription repression: a genome-wide analysis. *Mol. Biol. Cell.* **16**, 2903–2912.

3. Chauhan, N., Inglis, D., Roman, E., Pla, J., Calera, J.A., and Calderone, R. (2003) *Candida albicans* response regulator gene *SSK1* regulates a subset of genes whose functions are associated with cell wall biosynthesis and adaptation to oxidative stress. *Eukaryot. Cell* **2**, 1018–1024.

4. Fradin, C., Kretschmar, M., Nichterlein, T., Gaillardin, C., d'Enfert, C., and Hube, B. (2003) Stage-specific gene expression of *Candida albicans* in human blood. *Mol. Microbiol.* **47**, 1523–1543.

5. Enjalbert, B., Nantel, A., and Whiteway, M. (2003) Stress-induced gene expression in *Candida albicans*: absence of a general stress response. *Mol. Biol. Cell* **14**, 1460–1467.

6. Nantel, A., Dignard, D., Bachewich, C., Harcus, D., Marcil, A., Bouin, A. P., Sensen, C. W., Hogues, H., van het Hoog, M., Gordon, P., Rigby, T., Benoit, F., Tessier, D. C., Thomas, D. Y., and Whiteway, M. (2002) Transcription profiling of *Candida albicans* cells undergoing the yeast-to-hyphal transition. *Mol. Biol Cell* **13**, 3452–3465.

7. De Backer, M. D., Ilyina, T., Ma, X. J., Vandoninck, S., Luyten, W. H., and Vanden Bossche, H. (2001) Genomic profiling of the response of *Candida albicans* to intraconazole treatment using a DNA microarray. *Antimicrob. Agents Chemother.* **45**, 1660–1667.

8. Fradin, C., De Groot, P., MacCallum, D., Schaller, M., Klis, F., Odds, F. C., and Hube, B. (2005) Granulocytes govern the transcriptional response, morphology and proliferation of *Candida albicans* in human blood. *Mol. Microbiol.* **56**, 397–415.

9. Ruben-Bejerano, I., Fraser, I., Grisafi, P., and Fink, G. R. (2003) Phagocytosis by neutrophils induces amino acid deprivation response in *Saccharomyces cerevisiae* and *Candida albicans*. *Proc. Nat. Acad. Sci. USA* **100**, 11007–11012.

10. Lorenz, M. C., Bender, J. A., and Fink, G. R. (2004) Transcriptional response of *Candida albicans* upon internalization by macrophages. *Eukaryot. Cell* **3**, 1076–1087.

11. Singh, V., Sinha, I., and Sadhale, P. P. (2005) Global analysis of altered gene expression during morphogenesis of *Candida albicans in vitro*. *Biochem. Biophys. Res. Comm.* **334**, 1149–1158.

12. Andes, D., Lepak, A., Pitula, A., Marchillo, K., and Clark, J. (2005) A simple approach for estimating gene expression in *Candida albicans* directly from a systemic infection site. *J. Infect. Dis.* **192**, 893–900.

13. Collart, M. A. and Oliviero, S. (1993) Preparation of Yeast RNA. In: *Current Protocols in Molecular Biology* (Ausubel, F. A., Kingston, R. E., Moore, D. D., Seidman, J. G., Smith, J. A. and Struhl, K. E., eds.), John Wiley and Sons, New York, pp. 13.12.1–13.12.5.

Chapter 10

In Vitro and *Ex Vivo* Assays of Virulence in *Candida albicans*

Richard A. Calderone

Abstract

The measurement of virulence using *ex vivo* and *in vitro* models is discussed in the context of the human pathogenic yeast, *Candida albicans*. The models described are of two types. First, reconstituted tissues of various sorts are used that are derived from human carcinomas. The tissues are grown *in vitro* in complex media, attain a three-dimensional tissue structure, and retain cell-surface antigens typical of the specific tissue. Both adherence and invasion of tissues can be studied following infection with strains of *C. albicans* (1, 2). Further, one can increase the level of complexity by providing infected tissues with host phagocytes or cytokines such that an immune contribution to protection can be followed (3–5). The second model employs *Drosophila melanogaster* larvae that are infected with *C. albicans* (6). In this model, the progression of virulence is followed after injection of strains of a pathogen of interest into the fly abdomen. Thus, in the case of human pathogenic fungi, the recognition of host tissues and invasion by the specific pathogen can be studied *in vitro* and correlations developed for human disease. The obvious advantage to using animal models (e.g., mice) is reduced cost, such that large numbers of *C. albicans* strains can be assessed for their virulence properties. Additionally, another application of these models is in drug discovery. It is clear that there are both advantages and disadvantages of the use of alternate models other than a murine model, to evaluate disease, and this is discussed below.

Key words: Virulence assessment, nonanimal models, *Drosophila*.

1. Introduction

First, I will suggest my definitions of *in vivo*, *ex vivo*, or *in vitro* models that are used to evaluate microbial virulence. Also, a relatively new application to the study of microbial virulence has been reported (see below) using the fruit fly, *Drosophila melanogaster*, termed a mini-host model, in obvious reference to the difference in size among this creature versus a mouse (6). That designation will be used in this chapter. An *ex vivo* assay usually refers to the removal of cells, from a test animal after an infection, e.g., alveolar macrophages following intranasal infection for various time periods with *Aspergillus fumigatus* conidia *(7)*. In this example, virulence/survival of the test organism is then determined by culturing lung lavage fluids on media that support the growth of the fungus and measuring the number of viable fungi recovered from the animals. The viability of a set of strains (and the killing of infected mice) in which a gene-deleted mutant is compared to other strains (wild-type and gene-reconstituted strains) can be compared for virulence differences, drug treatments, and drug efficacy, or even immune responses. In the latter instance, one can compare virulence in immune knockout mice or mice treated in ways to depress or activate specific arms of the immune response. This information is valuable to understand broad concepts of immunity or even to identify the activity of specific host cell populations, each through the use of "knockout" mice. Corticosteroid-treated animals, for example, were infected intranasally with *A. fumigatus*, a respiratory pathogen, to determine the consequence of that treatment with fungicidal activity of alveolar macrophages *(7, 8)*. So, herein, as the examples above describe, *ex vivo* refers to studies in which host cells are removed following infection of an animal host, and the particular model described above provides a way to evaluate virulence of fungal strains that infect the alveoli following their inhalation.

In vivo models are easy to define: animals are infected and assessed for the same types of endpoints just described, only always within the context of the animal. *In vitro* models are defined as those in which the host (be it an animal or a cell-line) is infected and that interaction is then studied in the context of a nonhost environment. In summary, the definitions of these types of models reflect where the interaction is studied i.e., in infected animals, in cells removed from infected animals, or in a host–fungus relationship strictly investigated in the laboratory.

The selection of an *in vitro*, *in vivo*, or *ex vivo* assay to evaluate some parameter reflects two primary considerations – cost and time – both of which are related. For example, to evaluate drug efficacy in an animal model requires a dose–response determination for the drug under study. In addition, the response of different pathogens, size of inocula, and the required controls must all be investigated. Such experiments become more complicated when additional parameters are evaluated, such as how the drug is delivered (i.v., peritoneal, intrathecial, etc.) and in what form (e.g., nonencapsulated or liposomal encapsulation). The number of mice required for such a study approaches several hundred, at a significant financial burden to the investigator (mice, housing per diem, salary of the investigators, etc.). Further, time is money. Of the many invasive animal models of candidiasis, the most commonly used model is infection through tail-vein injection, a time-consuming procedure. Ethical considerations also play an important role. Must animals die? Procedures need to be included in protocols that will minimize pain or alleviate pain. Animal death is not acceptable as an endpoint to an experiment; so increased surveillance of infected animals must be undertaken and parameters established to assess morbidity. The oversight provided by institutional reviews of protocols is absolutely essential. In addition, an animal model must mimic as closely as possible the disease in humans, a requirement that often goes unfulfilled. Likewise, *in vitro* models have their limitations since data from such studies only suggest correlations with disease. Infection and virulence of a selected mutant of *C. albicans* in a fruit fly model or in reconstituted tissue *in vitro* fits with candidiasis in a patient only as much as can be suggested by the experimental conditions. Attempts to circumvent this problem by creating a "host-like" environment include the addition of multiple host components such as serum to phagocytosis assays, the addition of phagocytes or other immune cells to studies of interactions *in vitro* of the fungus with cultured tissue, among other possibilities *(3–6)*. Two *in vitro* models are described below, one using cultured tissue derived from human carcinoma cells and the other using *D. melanogaster*.

2. Materials

2.1. In Vitro *Tissue Models*

1. *C. albicans* strains SC5314, CAF2-1, CAI4, or stains appropriate to the issue under investigation (*see* **Note 1**).
2. Human reconstituted tissues, Skinethic Laboratory, Nice, France (*see* **Note 2**).
3. Culture medium, brain heart infusion agar, 35 g/L, prepared one-day ahead of the tissue assay, stored at 4°C until needed.
4. Washing buffer, phosphate-buffered saline (PBS): 0.01 M phosphate, containing 0.86% NaCl, final pH 7.2.
5. Fixation of infected tissues. At appropriate time intervals, tissues and filters are removed from the microtiter plates, washed with PBS, and then prepared for fixation: 10% formalin in PBS. Following dehydration, tissue-filters are embedded in paraffin at room temperature. Sections of 4 µm are obtained, transferred to glass slides, and stained with periodic acid-Schiff (PAS) stain (*see* **Note 3**).
6. Scanning electron microscopy. PBS to remove nonadhering organisms; fixed with 2.5% glutaraldehyde and 2% formaldehyde, washed three times with PBS, postfixed with osmium tetroxide, washed again similarly with PBS, dehydrated in a graded series of ethanol solutions, and gold-coated after critical point drying.

2.2. In Vitro Tissue Models: Transcription Studies

1. RNA isolation. Infected or uninfected epithelial tissues are shock-frozen and total RNA is isolated using RNAPureTM (Peqlab, Erlangen, Germany). A number of similar kits are available for use.
2. cDNA synthesis is done using Superscript reverse transcriptase (Gibco, but this enzyme can be obtained from a number of distributors).
3. Quantitative RT-PCR. cDNAof 20 ng, amplification in real time in a Light-Cycler SYBR Green I kit (Roche), containing 3 mM Mg^{2+}, final concentration. Primer pairs, annealing temperature, and elongation time is optimized for each primer pair (*see* **Note 4**).
4. TES-phenol buffer for extracting RNA. 10 mM Tris-HCL, pH &.5, 10 mM 0.5 M EDTA, 0.5% SDS in DEPC H_2O, and 400 µL of phenol, pH 4.5.

2.3. Drosophila melanogaster – A Mimi-Host Model for Testing Virulence and Antifungal Drug Efficacy

1. *D. melanogaster* stocks – 2–4-day-old female, 30 flies used per treatment.
2. Yeast cells – overnight cultures in yeast-peptone-dextrose (YPD) medium.
3. Infections are initiated by injecting 1×10^7–1×10^{10} yeast cells into the thorax of individual flies.

3. Methods

It is important to realize that measuring the amount of organism adhered to tissues in this system is only semiquantitative. Visualization of adhering organism is accomplished by light microscopy of fixed/stained tissue sections or by scanning electron microscopy. Nevertheless, general conclusions can be made on the order of events in timed experiments for specific sets of mutants. For example, wild-type strains were observed more adherent than two-component signal transduction mutants as well as secreted aspartyl protease (SAP) mutants over a period of 24 h *(1, 2, 5)*. In addition, strains can be evaluated for penetration into the tissue by following the infection for longer periods of time (24–48 h). The other major observations that can be measured include host responses, i.e., cytokines, β-defensins, or in the case of fungal products, SAP antigen or transcript *(5)*. The tissue model allows the investigator to broaden the objectives of the experiments for either host or fungus.

3.1. In Vitro Tissue Models

C albicans SC5314 or CAF2-1 or other suitable strains as the specific investigation dictates can be used. As mentioned above, the model is useful for evaluating the virulence of any single gene-deleted strain and its set of controls. In the case of *C. albicans*, strain sets need to include a single gene-deleted strain as well as the gene-reconstituted strain in addition to the companion null strain lacking both genes. As the selection of correct strains is done via Ura3⁻ conversion to prototrophy, then loss of *URA3* in transformants provides a selection for the second gene knockout; *URA3* positional affects have been described that may influence phenotypes since the *URA3* gene is not in its native locus *(9)* (*see* **Note 5**).

1. Wild-type strains are grown in one's favorite medium (BHI or yeast-peptone-glucose are the most common), and then adjusted to an inoculum of 10^6–10^7 cells by hemocytometer counts. This somewhat tedious method can be bypassed by instead using a cell concentration that corresponds to a known optical density reading.
2. There are several types of reconstituted tissues provided by Skinethic Laboratory, Nice, France, that are of either epidermis or endodermis origins. Most of the tissues are derived from human cancer cell lines. The tissues are provided on polycarbonate filters in a microtiter well format that is also available as a 96-well high throughput screen (HTS) that may be useful for antifungal drug testing, although this objective has not been reported as yet.
3. The reconstituted tissue must be cultured for at least 24–72 h after arrival to the lab. For the reconstituted human esophageal tissue, the company provides a chemically defined medium (MCDB 153 medium, containing gentamicin (25 µg/mL), 5 µg/mL of insulin, and 1.5 mM Ca^{2+}) for growth and maintenance of the tissue. This medium is sufficient to allow additional growth of tissues as needed. Instructions from the producer are well-defined and help in planning the timing of experiments. Some variation in the medium may reflect growth requirements of different tissues. For inoculation, protocols vary in regard to inoculum size, as indicated above, but tissues (per well) can be infected in a volume of 50 µL of PBS, pH 7.2.
4. For measuring the amount of organism adhered to tissues at designated time points, the infected tissues are washed three times with PBS and fixed in 10% formalin in PBS for at least 24 h.
5. Fixed cells are then dehydrated, sectioned, the sections transferred to glass slides, and stained with the PAS. Light microscopic visualization allows for qualitative inferences in the amount of organism adhering to tissues, and at later times, the invasion of the tissue can be analyzed.
6. Scanning EM allows for an analysis of interactions at the surface of the tissues and at greater magnification. Such analysis can also be useful in antigen tagging experiments, as for example those reported for Sap proteins *(2)*. Tissue damage caused by the organism can be measured by the release of lactate

dehydrogenase using spectrophotometric measurements of NADH oxidation following the conversion of pyruvate to lactate. Several assays are available, e.g., see the Wroblewski–La Due method *(2)*.

3.2. In Vitro *Tissue Models: Transcription Studies*

1. In these experiments, measurements of tissue responses that are associated with immunity to the organism can be accomplished. For instance, the interaction of *C. albicans* with oral epithelial tissue was studied in order to understand the role of the cytokine interleukin-8 (IL-8) in protection against candidiasis *(3)*. The reasoning is that IL-8 provides an important role in T-helper 1 responses, which is critical to protection in oral candidiasis. In these experiments, epithelial tissues were infected with a specific inoculum of washed yeast cells of several species of *Candida*. At various times postinfection, infected or uninfected tissues were collected and centrifuged. The amount of secreted IL-8 in the supernatant of the infected tissues was detected using a sandwich ELISA method with optimized amounts of a mouse antihuman IL-8 monoclonal antibody (G265-8, PharMingen, San Diego, CA, USA). For assays, this antibody is coated in microtiter wells overnight, washed, nonspecific binding prevented, the tissue-fungus supernatants are added, and allowed to incubate overnight. Then, the primary antibody supernatants are washed several times with PBS, and a biotin-labeled anti-IL-8 monoclonal is added. Again washing is critical to remove nonbound antibody and the amount of IL-8 is determined using an avidin-biotin-alkaline phosphatase complex (DAKO) followed by the addition of the phosphatase substrate that is compared to assays with increasing concentrations of IL-8. Optical densities at 405 nm are obtained from experimental samples and the amount of IL-8 determined from the linear range of standard curves with the recombinant IL-8. The authors were able to correlate tissue damage by *C. albicans* with increased responses of IL-8, while *C. glabrata* and *C. tropicalis* produced less damage and also less IL-8. Heat-killed *C. albicans* failed to induce an IL-8 response.
2. cDNA synthesis is performed using standard methods as described above in **Section 2.2**.
3. RT-PCR. Infected tissues are processed as described next *(5)*.

   Each tissue, infected or otherwise, is transferred to RNase/DNase-treated 1.5 mL microcentrifuge tubes containing a standard buffer. The samples are centrifuged at 13,000 rpm for 1 min and the supernatant decanted, tissues are suspended in 400 µL of TES (10 mM Tris·HCl pH 7.5, 10 mM 0.5 M EDTA, 0.5% SDS; in DEPC H_2O) and 400 µL of phenol pH 4.5. After vortexing for 10 s, the samples are incubated at 67°C for 1 h, vortexing every 10–15 min, placed on ice for 5 min, and then centrifuged for

5 min at 13,000 rpm. The top layer is transferred to a 1.5-mL microcentrifuge tube and 400 μL of phenol (pH 4.5) added. The mixture is vortexed and incubated on ice for 5 min. After centrifuging for 5 min at 13,000 rpm, the top layer is transferred to a new 1.5-mL microcentrifuge tube, and 400 μL of chloroform:isoamyl alcohol (24:1) is added, vortexed, and placed on ice for 5 min. After vortexing and centrifuging as above, the top layer is transferred to a new 1.5-mL microcentrifuge tube. DEPC-treated sodium acetate of 40 μL and 1 mL of ice-cold 100% ethanol are added to the mixtures, which are again vortexed and incubated on ice for 5 min. After centrifugation for 5 min at 13,000 rpm, the supernatant is removed and 500 μL of ice-cold 70% ethanol is added. Again, after centrifugation, the supernatant is discarded, the pellet air-dried, and suspended in 100 μL DEPC H_2O. The concentration and purity of the RNA was determined at wavelengths of 260 and 280 nm using a spectrophotometer.

4. Quantification. The RT-PCR reactions to quantify gene transcription are performed according to the protocol for the Qiagen One-Step RT-PCR Kit.

    (a) A 50-μL reaction mixture included 25 μL of RNase-treated H_2O, 10 μL of 5X buffer, 2 μL of dNTP mix (10 mM each), 0.5 μg each of primers Ssk2 NB 5′ and Ssk2 NB 3′, 1 μg of RNA, and 2 μL of enzyme mix.

    (b) The reactions are set up in duplicate to detect both transcripts for all samples. The amplification conditions for detecting the transcript can vary by individual design but may follow this program: 30 min at 50°C (reverse transcription), 15 min at 95°C (activation of the HotStarTaq Polymerase), followed by 25 cycles of 1 min at 94°C, 30 s at 56°C, and 55 s at 72°C, concluding with a final extension step of 10 min at 72°C.

    (c) The amplification conditions for detecting the *ACT1* transcript (internal control) are as described above, except annealing is performed for 30 s at 59°C and extension for 45 s at 72°C. A gel imager (Alpha Imager 2000, Alpha Innotech Corp.) is used to quantify the intensity of transcripts.

    (d) The levels of transcripts are expressed as a ratio and calculated by dividing the band intensity of each gene by the band intensity of *ACT1* for each RNA sample. A fold-difference is determined and calculations are performed for all six samples and three independent experiments. The two-tailed Student's *t*-test was utilized to determine the significance of the values.

3.3. *D. melanogaster* Model: Inoculation

1. Oregon R flies are used as wt flies. Manipulation, feeding, and housing of flies is described *(10)*.
2. Yeast cells are grown and maintained on YPD agar.

3. Inoculation into the *thorax* is done using a thin, sterile needle (0.25 mm diameter) dipped into a suspension of 1×10^7–1×10^{10}, and it is then used to inject the organism into the thorax. Alternatively, for ingestion assays, fly-food vials containing YPD agar on which 1×10^7 cells are grown for 24 h may be used. Flies are fed for 48 h, and then switched to fly-food and maintained at 28°C. After inoculation (*see* **Section 3.3.3**), flies are maintained at 29°C. If death occurs within 3 days, such flies should not be included in data analysis. During the infection process, flies are changed to fresh vials containing food every 2 days. Survival is assessed daily for a total of 8 days (30 flies per strain). Quantification of the organism used for inoculation is determined by dipping the inoculum needle used for infection in sterile saline and determining colony-forming unit (CFU) by plating dilutions of this suspension onto YPD plates, which were incubated at 30°C for 48 h. Likewise, for the ingestion model, flies are evaluated for survival, daily for 8 days. For measurements of tissue burden, groups of 20 flies are collected a 0, 12, 24, and 36 h after infection, then transferred to plastic tubes, and ground in 1.0 mL of physiological saline. Tissue loads of the organism are determined by plating aliquots on YPD medium and CFU determined as described above.

Histopathological assessment of infection is determined as follows: following either type of assay (ingestion or thorax inoculation), at day 2 postinfection, flies are fixed with 10% (vol/vol) formaldehyde, stained with Grocott-Gomori methenamine silver nitrate, and sections examined for fungal growth by microscopy.

The assay can also be evaluated for drug-efficacy testing using antifungals such as fluconazole. In these experiments, flies are first starved for 6–8 h and then transferred to vials containing their food supplemented with fluconazole (1 mg/mL). After 24 h, flies are infected by injection as described above, and then transferred to new vials containing fluconazole. Control flies (uninfected) are treated similarly but not fed with fluconazole. Fluconazole protection was monitored daily for 8 days.

4. Notes

1. Strain CAI4 is Uri⁻ and, therefore, is not used in comparison to gene-deleted strains in which *URA3* has been restored either to the locus of the gene that was deleted or its own locus.
2. Tissues that are infected should be evaluated for adherence and invasion within a reasonable amount of time (0–4 days) since, especially in wt strains germination and growth of the organism occur quickly and overwhelm the microscopic fields with extensive hyphal development.

3. We find that the PAS stain gives much better contrast in microscopic examination; the fungus retains a reddish color against the colorless background.
4. Real-time PCR is preferred to the standard RT-PCR that is more often reported. Real-time PCR allows one to evaluate transcription at different times within the PCR amplification steps.
5. Ura3-positional affects on phenotype are reported using the Urablaster method to obtain specific gene-deleted mutants. This problem can often present itself if phenotypes do not match gene dosage. For example, a single gene-carrying strain should have a phenotype that matches a single gene copy compared to wet strains, which have two copies of the specific gene.

References

1. Bernhardt, J., Herman, D., Sheridan, M., and Calderone, R. (2001) Adherence and invasion studies of *Candida albicans* strains, using *in vitro* models of esophageal candidiasis. *J. Infect. Dis.* **184**, 1170–1175.
2. Li, D., Sheridan, M., and Calderone, R. (2002). Temporal expression of the *Candida albicans* genes *CHK1* and *SSK1*, adherence, and morphogenesis in a model of reconstituted human esophageal epithelia candidiasis. *Infect. Immun.* **70**, 1558–1565.
3. Schaller, M., Mailhammer, R., and Korting, H. (2002) Cytokine expression induced by *Candida albicans* in a model of cutaneous candidosis based upon reconstituted human epidermis. *J. Med. Microbiol.* **51**, 672–676.
4. Schaller, M., Mailhammer, M., Grassl, G., Sander, C., Hube, B., and Korting, H. (2002) Infection of human oral epithelia with *Candida* species induces cytokine expression correlated to the degree of virulence. *J. Invest. Dermatol.* **118**, 652–657.
5. Schaller, M., Bein, M., Korting, H., Baur, S., Hamm, Monod, M., et al. (2003) The secreted aspartyl proteinases Sap1 and Sap2 cause tissue damage in an in vitro model of vaginal candidiasis based on reconstituted human vaginal epithelium. *Infect. Immun.* **71**, 3227–3234.
6. Chamilos, G., Lionakis, M., Lewis, R. E., Lopez-Ribot, J., Saville, S., Albert, N., Halder, G., and Kontoyiannis, D. (2006) *Drosophila melanogaster* as a facile model for large-scale studies of virulence mechanisms and antifungal drug efficacy in *Candida* species. *J. Infect. Dis.* **193**, 1014–1022.
7. Latge, J.-P. (2001) The pathobiology of *Aspergillus fumigatus*. *Trends Microbiol.* **9**, 382–389.
8. Hachem, R., Baha, P., Hanna, H., Stephens, L. C., and Raad, I. (2006). EDTA as an adjunct agent for invasive pulmonary aspergillosis in a rodent model. *Antimicrob. Ag. Chemother.* **50**, 1823–1827.
9. Cheng, S., Nguyen, M. H., Zhang, Z., Jia, H., Hanfield M., and Clancy, C. J. (2003). Evaluation of the role of four *Candida albicans* genes in virulence by using four gene disruption strains that express *URA3* from the native locus. *Infect. Immun.* **71**, 6101–6103.
10. Lemaitre, B., Nicolas, E., Michaut, L., Reichart, J. M., and Hoffmann, I. A. (1996) The dorsoventral regulatory gene cassette spatzle/Toll/cactus controls the potent antifungal response in *Drosophila* adults. *Cell* **86**, 973–983.

Part IV
Strain Typing and Identification

Chapter 11

Biotyping of *Candida albicans* and Other Fungi by Yeast Killer Toxins Sensitivity

Luciano Polonelli and Stefania Conti

Abstract

Intraspecific differentiation of pathogenic microorganisms is a major need in epidemiological studies concerning the source and spread of infections. This requirement is paramount for those etiologic agents of infectious diseases, which are mainly grouped into one species within the genus, such as *Candida albicans*. Ideally, laboratory methods for biotyping purposes should be sensitive, reproducible, easy, and economical to perform. In addition, the methods should be flexible in their application to taxonomically unrelated pathogens. We have shown that the toxins produced by a selected panel of killer yeasts, each characterized by a wide spectrum of antimicrobial activity, may be used to discriminate strains belonging to the species of the genus *Candida* and to other species of eukaryotic and prokaryotic pathogenic microorganisms. The "yeast killer system," which may be sharply increased in sensitivity by addition of further standardized killer yeasts, has proven to be of value in the resolution of many cases of clinical and nosocomial fungal infections. Owing to its reliability, economy, and versatility, this phenotypic system can be used as an alternative biotyping method in laboratories lacking the financial and training resources necessary to perform more sophisticated and expensive molecular approaches.

Key words: Killer yeasts, killer toxins, *Candida*, fungi, biotyping, epidemiology, infections.

1. Introduction

1.1. Background

The terminology defining intraspecific differentiation in microbiology is particularly poor and often confusing. The categories of subspecies, variety, subvariety, form, and biotype are bound to concepts of taxonomy, but the same definitions do not exactly coincide with those of epidemiology (1). The most common term used by microbiologists for intraspecific differentiation is "strain," while "biotype" is primarily used by epidemiologists interested in the correlation between individual strains, their habitats, and mode of transmission. An impelling condition for an effective biotyping system is that it is independent of other unrelated identification tests. This is necessary so that cumulative information may be acquired by a reasonable compromise between test simplicity and effectiveness in discriminating among the largest possible number of isolates to be processed.

The intraspecific differentiation of pathogenic microorganisms for epidemiological purposes has become a pressing need in the last few decades, particularly as a result of the increasing incidence of nosocomial infections and infections associated with acquired immunodeficiency syndrome (AIDS). Different phenotypic laboratory methods based on morphological, physiological, and serological criteria that may enable microbiologists to characterize single microbial strains and establish relationships between biotypes, habitats, patients, clinical cases, source, and spread of infections have been used beyond the classic sexual mating procedures *(1–8)*. More recently, genotypic methods have been proven very useful for strain differentiation. These molecular approaches, however, require the availability of proper facilities and technologies uncommon in small laboratories. Further phenotypic typing systems, not based on morphology or serotyping, have been developed for different microbial species. These take advantage of other factors such as phage susceptibility or enzyme production *(9–13)*. However, it is often difficult to use these methods with a large number of isolates or in small microbiological laboratories.

The adoption of one among the different methods depends on the particular requirements of the operator with regard to discriminative capacity, technical feasibility, scientific background, financial resources, time availability, and the number of microorganisms to be investigated.

In bacteriology, reliable biotyping systems for clinical isolates have been developed, based on the differential susceptibility to (or production of) bacteriocins, proteins produced in particular by Gram-negative bacteria, which are lethal to strains mostly belonging to the same species *(14–16)*.

A similar "killer phenomenon" has been recognized to occur among yeasts that relies on the production by some strains (killer yeasts) of (glyco)proteins (killer toxins), which are inhibitory to members (sensitive yeasts) of the same or other species *(17)*.

1.2. The Yeast Killer Phenomenon

The yeast killer phenomenon was first reported in 1963 by Bevan and Makower who described that a few isolates of *Saccharomyces cerevisiae* produced a substance lethal to other strains of the same species *(18)*. This effect was defined the killer phenomenon and the substance was named the *killer toxin*. Since that time, even though *S. cerevisiae* has represented the most investigated killer system, numerous studies have been performed to determine the distribution of the killer character among yeasts of human and natural origin, the modalities of their action, as well as the physiological and chemical properties of the toxins. As it is beyond the scope of this chapter to discuss such mechanisms in detail, the interested reader can find such descriptions elsewhere *(19–32)*.

1.3. The Yeast Killer System

The potential of the yeast killer phenomenon was first applied to the differentiation of strains within the same species of pathogenic yeasts. The original model investigated was *C. albicans* as the sensitive yeast and other genera and species of yeasts as killer strains. A wide range of killer yeasts, obtained from private and public fungal collections, was selected for the standardization and the potential increasing of the discriminatory power of the killer system (**Table 11.1**). Some of these species, particularly the ones once belonging to the genus *Hansenula* and currently grouped

Table 11.1
List of potential killer yeasts selected for biotyping purposes

Strain	Collection	No
Pichia spp.	Stumm	1034
Pichia spp.	Stumm	1035
P. anomala	UM	3
P. anomala	CBS	5739
P. anomala	Ahearn	Um866
P. californica	Ahearn	WC40
P. canadensis	Ahearn	WC41
P. dimennae	Ahearn	WC44
Williopsis mrakii	Ahearn	WC51
P. kluyveri	Stumm	1002
P. anomala	UT	12
P. bimundalis	Ahearn	WC38
P. fabianii	CBS	5640
P. petersonii	Ahearn	WC53

Table 11.1 (continued)

Strain	Collection	No
P. guilliermondii	UT	19
S. cerevisiae	CDC	B2210
P. bimundalis	CBS	5642
P. fabianii	Ahearn	WC45
P. holstii	CBS	4140
P. subpelliculosa	CBS	5767
P. omheri	UCSC	0
Candida guilliermondii	UCSC	0
C. maltosa	UCSC	0
P. spartiniae	UCSC	0
P. nonfermentans	UM	200
P. carsonii	CBS	810
P. farinosa	CBS	185
P. guilliermondii	CBS	2031
C. kefyr	UP	241
C. kefyr	UP	254
C. kefyr	UP	330
P. kluyveri	UP	5F
P. kluyveri	UP	6F
P. membranaefaciens	UP	10F
P. kluyveri	UP	11F
P. anomala	ATCC	96603
W. mrakii	UCSC	255
S. cerevisiae	Kandel	SC5
S. cerevisiae	Kandel	SC8
Kluyveromyces lactis	Gunge	NK-1
K. lactis	Gunge	2105-ID

Table 11.1 (continued)

Strain	Collection	No
K. lactis	Gunge	IFO 1267
P. americana	NRRL	Y-2156
P. angusta	NRRL	Y-2214
P. anomala	NRRL	Y-366
P. ciferrii	NRRL	Y-1031
P. dryadoides	NRRL	Y-10990
P. euphorbiaphila	NRRL	Y-12742
P. lynferdii	NRRL	Y-7723
P. muscicola	NRRL	Y-7005
P. petersonii	NRRL	YB-3808
P. populi	NRRL	Y-12728
P. silvicola	NRRL	Y-1678

ATCC, American Type Culture Collection, USA; Ahearn, D.G. Ahearn, Georgia State University, Atlanta, GA, USA; CBS, Centraalbureau voor Schimmelcultures, The Netherlands; CDC, Centers for Disease Control, Atlanta, GA, USA; Gunge, N. Gunge, Kumamoto Institute of Technology, Kumamoto, Japan; Kandel, J. Kandel, California State University, Fullerton, CA, USA; NRRL, Northern Regional Research Laboratory, Peoria, IL, USA; Stumm, C. Stumm, University of Njimegen, Njimegen, The Netherlands; UCSC, Istituto di Microbiologia, Universitá Cattolica del Sacro Cuore, Rome, Italy; UM, University of Milan, Italy; UP, University of Parma, Italy; UT, University of Turin, Italy.

under the genera *Pichia* and *Williopsis* according to criteria of DNA hybridization, have proven to exert a microbicidal differential activity also against other taxonomically unrelated yeast, mold, and bacterial isolates *(33, 34)*.

The toxic effect of each killer yeast could be very simply visualized within a system standardized to define growth parameters either for the killer or the sensitive strain, such as medium composition, incubation temperature, inoculum size, test modalities, etc. The evaluation of the presence or lack of a halo of inhibition around each selected killer yeast streaked onto the surface of the test plate against the sensitive yeast integrated into the medium allowed establishment of the total spectrum of susceptibilities (**Fig. 11.1**). After standardization of the procedure, a simple "killer system" was developed that permitted the intraspecific differentiation of strains belonging to *C. albicans* as well as to other yeast, mold, and bacterial species *(35–42)*. The yeast killer

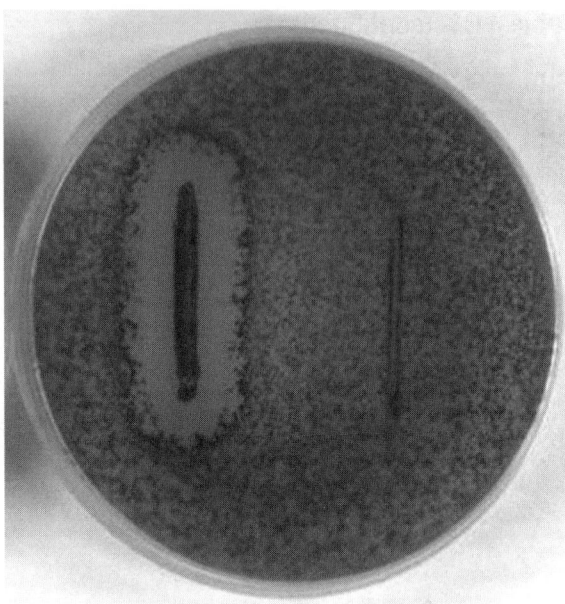

Fig. 11.1. Positive (*left*) and negative (*right*) killer effect among yeasts.

system has been diversified by using, in place of the producing killer strains, the respective purified killer toxins. This method was independent of all the different characteristics of growth of the concurring killer yeasts and allowed further standardization. The remarkable reproducibility of the behavior of selected yeast killer toxins tested at defined concentrations allowed use of the width of the halo of inhibition that had been produced in order to define the biotype of the investigated yeast-sensitive strain. The system based on purified yeast killer toxins has been also automated by application of computer-aided programs allowing to establish the range of probability of appurtenance of yeasts to reference biotypes according to a predefined margin of error (i.e., 5%) *(43)*. A modification of the killer technique, performed by incubating pure conidial and yeast cell inocula with killer toxins, could permit their activity to be evaluated by comparing the numbers of colony forming units (CFU) with those obtained by incubating the same inocula with a proper control *(44)*.

Critical reviews of different typing systems for *C. albicans* and their applications (epidemiologial research, investigation of outbreaks of diseases, study of virulence, etc.) including phenotypic methods such as antibiogram, serotyping, carbon assimilation pattern, morphotyping, resistotyping, biotyping, killer toxins typing, immunoblotting, isoenzyme analysis, as well as molecular approaches such as analysis of DNA restriction fragment length polymorphism, karyotyping; and the use of DNA probes proved that none of them is ideal *(45, 46)*. Phenotypically delineated strains usually do not correlate with their pathogenic potential although they may be useful in epidemiological investigations. Most of phenotypic and genotypic available typing methods, however, have not been adequately assessed by the scientific community, and some present only poor discriminatory power or reproducibility. As assessed before, a reasonable criterion might be the one of using in comparison different approaches for biotyping purposes of *C. albicans* and other pathogenic microorganisms. Overall, the yeast killer system, owing to its intrinsic characteristics of flexibility, easiness, economy, reproducibility, and sensitivity, may represent a possible unique choice for all those laboratories that are not familiar with more sophisticated molecular technologies for the strain differentiation of each group of microorganisms. Details concerning the use of such a system are detailed below.

2. Materials

It is important to point out that all buffers, reagents, solutions, strains, and any other material or reagent necessary to carry out the yeast killer system are not toxic and do not represent any radiation hazard. They do not require, moreover, particular conditions for maintenance of stability and effectiveness.

2.1. Killer Yeasts

Killer yeasts may be obtained from public and private collections (**Table 11.1**). Their selection is made according to their killer characteristics of reproducibility and heterogeneous behavior against a wide range of sensitive microorganisms properly tested *(34–42)* (*see* **Note 1**). The number of adopted killer yeasts determines the degree of sensitivity of the system.

2.2. Yeast and Molds Investigated by the Killer System

Eukaryotic microorganisms, isolated by clinical samples or obtained from private and public microbial collections, which have been investigated by the killer system, are reported below *(34–37, 40, 42–44, 47–58)*.

2.2.1. Yeasts

C. albicans, C. glabrata, C. kefyr (C. pseudotropicalis), C. tropicalis, Cryptococcus neoformans var. *neoformans, Cryp. neoformans* var. *grubii, Cryp. neoformans* var. *gattii, Debaryomyces hansenii, Kluyveromyces lactis, Malassezia furfur, M. pachydermatis, Pichia anomala, S. cerevisiae, S. exiguus, Zygosaccharomyces baillii.*

2.2.2. Molds

Aspergillus brevipes, A. flavus, A. fumigatus, A. fumigatus var. *acolumnaris, A. fumigatus* var. *ellipticus, A. fumigatus* var.

sclerotiorum, A. nidulans, A. niger, A. parasiticus, A. unilateralis, A. viridinutans, Aureobasidium pullulans, Cladophialophora bantiana, Cunninghamella elegans, Curvularia lunata, Exophiala jeanselmei, Fusarium sporotrichoides, F. chlamydosporum, F. tricinctum, F. solani, F. poae, Fonsecaea pedrosoi, Microsporum canis, Neosartorya aurata, N. aureola, N. fennelliae, N. fischeri var. *fischeri, N. fischeri* var. *glabra, N. fischeri* var. *spinosa, N. quadricincta, N. spathulata, Penicillium camemberti, P. melanochlorum, P. notatum, P. palitans, P. roqueforti, Phaeoanellomyces werneckii, Phialophora verrucosa, Pseudoallescheria boydii, Rhizopus microsporus, Scopulariopsis brevicaulis, Sporothrix schenckii* (mycelial and yeast form).

2.3. Media

2.3.1. Yeasts

1. The killer yeasts are cultured on Sabouraud dextrose agar (Difco Laboratories, USA) plates (pH 5.6).

2. Sabouraud dextrose broth (Difco) buffered at pH 3 or yeast–extract–peptone–dextrose (YEPD) (Difco) (2% peptone, 2% dextrose, 1% yeast extract) broth buffered at pH 4.5 with 0.1 M citric acid and 0.2 M sodium phosphate dibasic anhydrous are the media used for producing yeast killer toxins *(43)* (*see* **Note 2**).

3. The killing assay against yeasts is carried out by using YEPD agar buffered at pH 4.5 with 0.1 M citric acid and 0.2 M potassium phosphate dibasic anhydrous and added with methylene blue (final concentration 0.003%).

4. Commercially modified Sabouraud dextrose agar buffered at pH 4.5 with 0.1 M citric acid and 0.2 M potassium phosphate dibasic anhydrous and added with 0.003% methylene blue may be successfully used to replace YEPD agar *(35, 36)*.

5. Brain heart infusion agar (Difco) and blood agar base (Difco) supplemented with 5% defibrinated horse blood (Sclavo Laboratories, Italy) are used in the study of the activity of killer toxins against the yeast form of *S. schenkii*.

6. Sabouraud dextrose agar at pH 5.6 is used as the growth and maintenance medium for the sensitive yeasts isolates *(35, 36)*.

2.3.2. Molds

1. YEPD agar buffered at pH 4.5 with 0.1 M citric acid and 0.2 M potassium phosphate dibasic anhydrous added with methylene blue (0.003%).

2. Commercially modified Sabouraud dextrose agar buffered at pH 4.5 with 0.1 M citric acid and 0.2 M potassium phosphate

dibasic anhydrous and added with 0.003% methylene blue may be alternatively used.

3. Sabouraud dextrose agar at pH 5.6 is the medium used to maintain the fungus cultures to be tested by the killer system *(37, 40)*.

3. Methods (see Note 3)

3.1. Preparation of Yeast Killer Toxins (43, 44, 52) (see Note 4)

1. The yeast killer toxins are obtained from broth cultures of the producing killer yeasts grown in Sabouraud dextrose broth buffered at pH 3 or in YEPD broth buffered at pH 4.5 with 0.1 M citric acid and 0.2 M dibasic anhydrous sodium phosphate.

2. Suitable amounts of broth are inoculated with a loopful of killer yeast grown for 24 h in Sabouraud dextrose agar plates at 30°C and incubated at 25–30°C for 18 h/7 days with or without shaken conditions (70 rpm).

3. After this period, the cells are removed by centrifugation at $1500 \times g$ at 4°C for 10 min and the supernatant is filtered and concentrated $50 \times$ with a PM 10 membrane in a TCF10A ultrafiltration cell under N pressure (Amicon Corporation, USA) or by Minicon B15 concentrators.

4. The protein and carbohydrate contents of the concentrated product are determined by the Lowry method and Antrone test, respectively.

5. A volume of 4.5 mL of the concentrated killer toxins (approximately 40 mg protein) is loaded onto a Sephadex G50-fine column (80 cm × 2 cm dimension), previously equilibrated with a sodium acetate buffer, 1 mM, pH 4.5.

6. The column is calibrated by means of a set of proteins of known molecular weight: ovalbumin, chymotripsinogen a, and ribonuclease a (Sigma, USA). The detection system is a 2238 Uvicord SII (LKB Corp.) equipped with a flow cell and 278 filters.

7. The flow rate is 35 cm/h and fractions of 5 mL are collected *(43)*. The active fractions are selected for the killer system (*see* **Notes 5 and 6**).

3.2. Killer System for Yeast Strain Differentiation (35, 36)

1. The potential sensitive yeast isolates to be investigated are cultured on Sabouraud dextrose agar plates incubated for 24 h at 30°C.

2. A yeast suspension containing approximately 2×10^5 cells/mL is produced from a single isolated colony and incubated for

18 h at 25°C with shaking (120 rpm) in 10 mL of YEPD broth (pH 4.5).

3. One milliliter from each broth culture is diluted with 10 mL of fresh YEPD broth (pH 4.5).

4. One milliliter of this suspension (*see* **Note 7**) is then mixed with 20 mL of YEPD agar (pH 4.5) or modified Sabouraud dextrose agar (pH 4.5) containing 0.003% methylene blue kept melted at 45°C, and the mixture is poured into a Petri dish to obtain an agar–yeast suspension and allowed to solidify (*see* **Note 8**). The dye differentially stains the dead yeast cells blue.

5. The yeast killer strains are streaked onto the surface of the agar plates integrated or superficially inoculated with the potential sensitive yeast isolates to be tested (*see* **Note 9**).

6. For testing the activity of killer toxins, 10-mm-diameter wells are pouched in the agar and filled with 100 µL of (partially) purified killer fractions or the crude concentrated killer toxins.

7. The agar plates are then incubated at 25°C for 48–72 h (according to the growth rate of the yeast species to be investigated) before reading the results.

3.3. Killer System for Mold Strain Differentiation (37, 40, 42)

1. Heavy conidial suspensions are obtained by gently scraping in distilled sterile water from the mycelial cultures of the moulds to be tested, which have been grown in Sabouraud dextrose agar slants (pH 5.6) or Potato dextrose agar (Difco) incubated for 7–14 days at 25°C.

2. The suspensions usually contain mostly conidia and are free of massive hyphal fragments as determined by light microscopy.

3. The number of conidia is standardized to an optical density of 1 on the McFarland standard (*see* **Note 10**).

4. One milliliter of such suspension is mixed with 20 mL of buffered YEPD or modified Sabouraud dextrose agar maintained at 45°C, and the mixture is poured into a Petri dish and let to solidify (*see* **Note 8**).

5. The killer yeasts are streaked onto the surface of the agar and the plates are incubated at 25°C until the growth of the mould to be tested is visible (*see* **Note 9**).

3.4. Modified Killer System (44)

In the study of susceptibility to killer toxins of the two forms (yeast and mycelial) of *S. scenckii*, a modified killer system, extensible to other yeasts and molds, is usable.

1. Yeast cells obtained by repeatedly subculturing the mycelial cultures of the fungus on blood agar plates incubated at 37°C under 5% CO_2 until yeastlike colonies develop, and conidial suspensions obtained from 15-day-old cultures grown on Sabouraud dextrose agar slant tubes are suspended in sterile distilled water. Optical microscopy is used to confirm that the yeast suspension contains only budding and nonbudding cells and the conidial suspension is devoid of hyphal fragments. The yeast and conidial concentrations are determined by counting in a Burker chamber and 20 µL of inocula are standardized to obtain an easily readable range of CFU.

2. Standardized suspensions are subjected to the action of 1 mL of selected yeast killer toxins. Each test is made in triplicate for statistical purposes.

3. The killer toxins are allowed to act overnight at 4°C to prevent conidial germination and yeast-cell budding or germ-tube-like formation.

4. After this period, the total amount (1 mL) of each suspension is mixed with 20 mL of melted Sabouraud dextrose agar for *S. schenckii* conidia, and brain heart infusion agar for *S. schenckii* yeast cells.

5. Control plates are prepared by inoculating Sabouraud dextrose agar (*S. schenckii* conidia inoculum) and brain heart infusion agar (*S. schenckii* yeast inoculum) plates with the same amounts of inocula diluted in Sabouraud dextrose broth (pH 4.5) and incubated overnight at 4°C.

6. The Sabouraud dextrose agar plates are incubated at 25°C and the brain heart infusion agar plates at 37°C under 5% CO_2 until yeast and mycelial colonies are clearly formed. This usually occurs after 3–4 days.

3.5. Reading and Interpretation of the Results

3.5.1. Killer Yeast Based System

1. The killer activity on fungal strains is considered positive when either a clear zone of inhibition or a region of blueish-colored cells, or both, surrounds the streaked killer yeast. A negative result is considered if neither of these effects is observed (**Figs. 11.2** and **11.3**).

2. For differentiating the strains within the species, the selected killer yeasts are grouped in triplets. Each triplet is examined for its pattern of activity against the microbial isolate to be investigated. The combined effect within each triplet is recorded according to conventional numerical codes, as borrowed from bacteriocin typing methods (**Table 11.2a**) or by conventional criteria of definition of microbial metabolic patterns (**Table** 11.2b) *(35–37, 40, 42)* (*see* **Note 11**). The sensitivity of the killer system may be implemented by using further triplets of killer yeasts previously tested in standardized conditions.

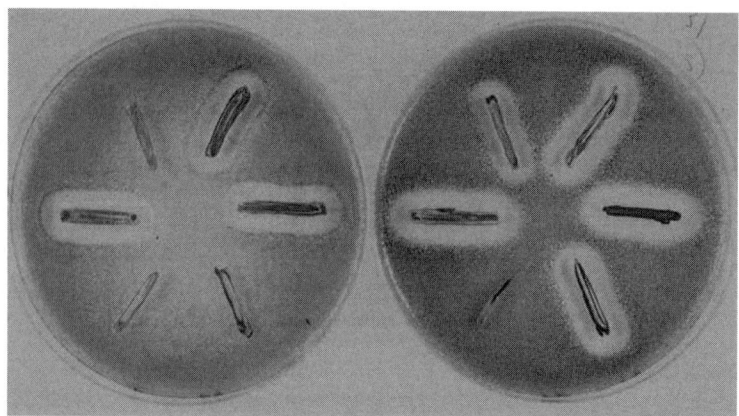

Fig. 11.2. Differential activities of streaked killer yeasts against two *C. albicans* isolates.

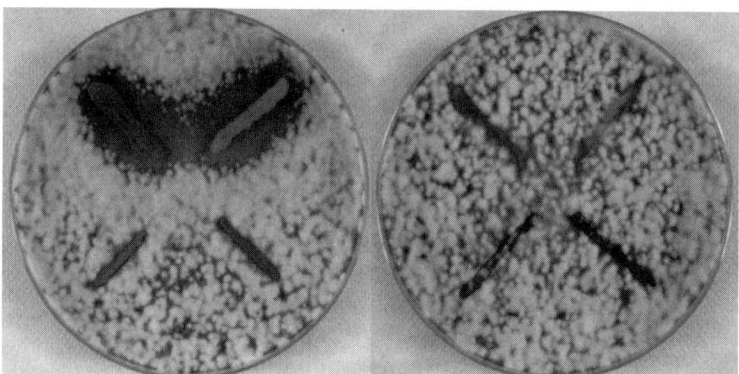

Fig. 11.3 Differential activities of streaked killer yeasts against two *Aspergillus fumigatus* isolates.

3.5.2. Killer Toxin Based System

1. In the killer system using killer toxins, the effect is considered positive either on the basis of a clear halo of inhibition or by the appearance of blue-tinged cells surrounding the well (**Fig. 11.4**). The activity is recorded as the inhibition diameter (ID) in millimeter. The treatment of the data may be aided by using a computer *(43)*.

2. The data are organized in a three-dimensional array of 100 × 8 × 2, where, for instance, 100 is the number of the yeast (*C. albicans*) isolates prejudicially investigated, 8 the number of the killer toxins used, and 2 represents the values obtained from duplicate analysis.

3. The two values are used for the computation of the between-assay standard deviation (O) and the coefficient of variation (CV) for any killer toxin response. The mean of the two values for all the sensitive yeast isolates for each killer toxin is represented on a histogram, where the abscissa is the value of the ID and the ordinate the number of yeast isolates with the same ID (**Fig. 11.5**).

Table 11.2
Numerical code of the activity of selected triplets of killer yeasts in the killer system

(a)

First killer activity	Second killer activity	Third killer activity	Code
+	+	+	1
+	+	−	2
+	−	+	3
−	+	+	4
+	−	−	5
−	+	−	6
−	−	+	7
−	−	−	8

(b)

First killer activity	Second killer activity	Third killer activity	Code
−	−	−	0
+	−	−	1
−	+	−	2
+	+	−	3
−	−	+	4
+	−	+	5
−	+	+	6
+	+	+	7

4. From the histogram and the value of the O for each killer toxin, appropriate ranges may be established, which allow division of *C. albicans* isolates into groups according to their susceptibility to the toxin. The combination of the groups obtained by the use of the eight killer toxins could provide a total number of biotypes equal to

Fig. 11.4. Differential activities of yeast killer toxins against two *C. albicans* isolates.

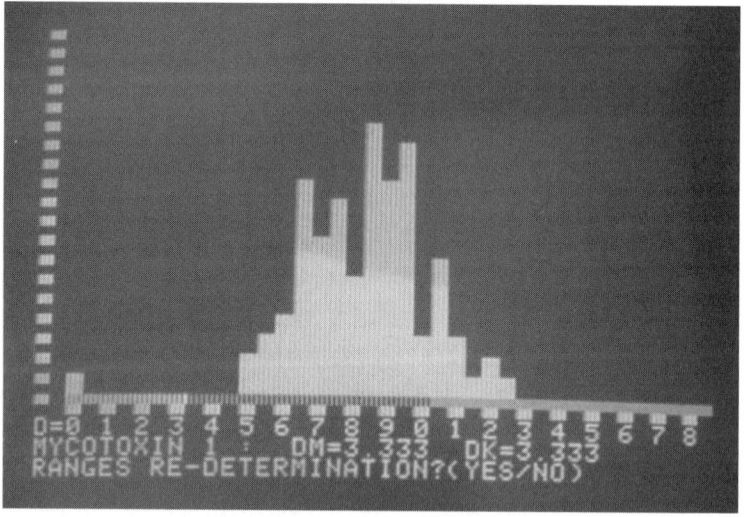

Fig. 11.5. Behavior of *C. albicans* isolates with respect to a killer toxin. The histogram reports in the abscissa the value of the IDs (mm) and in the ordinate the number of *C. albicans* with the same ID. (Reproduced with kind permission of Springer Science and Business Media.)

that obtained by multiplying the number of groups of each killer toxin by all the others (*see* **Note 12**). The biotype of the yeast strain investigated is referred to one or more groups. Each is expressed in terms of the percentage of probable affinity (**Fig. 11.6**) *(43)* (*see* **Note 11**).

3.5.3. Modified Killer System

The results are interpreted by counting the number of CFU on each plate inoculated with the numbered amount of yeast or conidial inocula subjected to the action of the selected killer toxins. The total number of expected colonies is assumed by counting the CFU on the control plates (**Fig. 11.7**). The percentage of inhibition of the investigated isolates determined by each

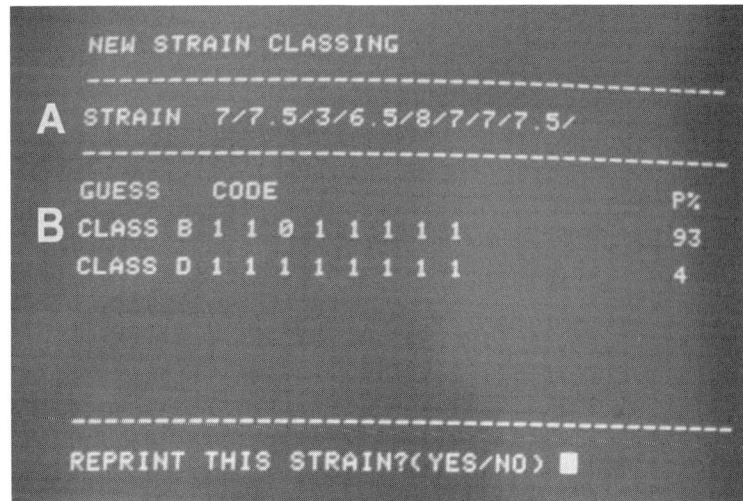

Fig. 11.6. Classification of a new *C. albicans* isolate with the computer-aided killer system. (**A**) ID with respect to eight killer toxins; (**B**) classes of affinity in percentage of the new isolate. (Reproduced with kind permission of Springer Science and Business Media.)

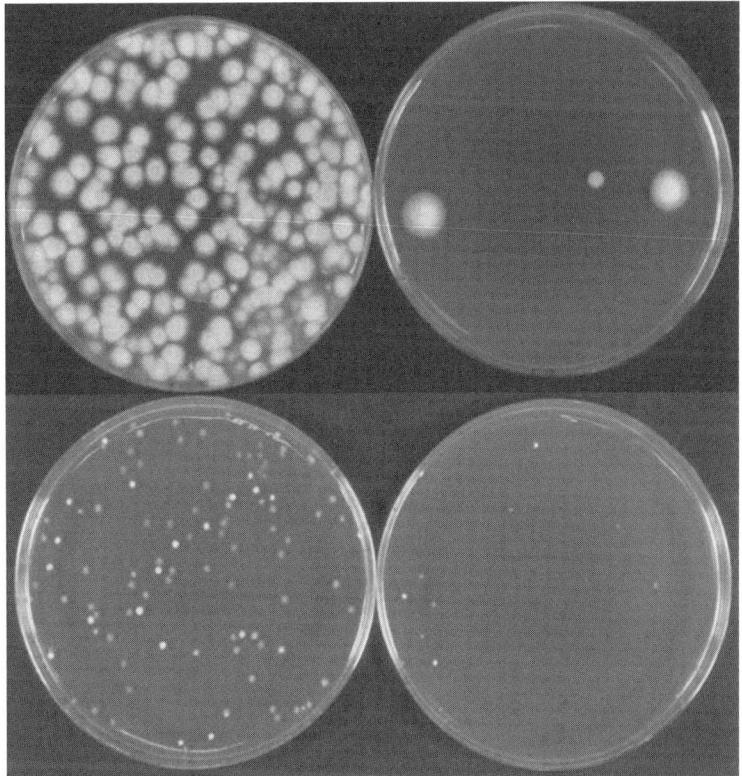

Fig. 11.7. Effect of a yeast killer toxin (*right plates*) on the number of CFU of a *S. schenckii* isolate mycelial (*top*) and yeast (*down*) forms. *Left plates*: control.

killer toxin may be used for the definition of biotypes according to established criteria (*see* **Note 11**).

4. Notes

1. By studying hundreds of cultures of killer yeasts, it has been observed that the highest effect is exerted by strains belonging to the genera *Pichia* and *Williopsis,* with a significant variability within each species *(36)*. One critical issue in the performance of the killer system may be the need to standardize and define outside any single laboratory the triplets of the killer yeasts to be used in order to make comparable the interlaboratory results.

2. The substitution of buffered, commercially modified Sabouraud dextrose broth for buffered YEPD broth does not significantly alter the results.

3. Temperature, pH, medium composition, and time of incubation have proven to be the most important factors affecting the activity of killer yeasts and toxins against sensitive microorganisms *(35, 43, 44)*. Standardized culture conditions should be adopted in order to obtain reproducible results even though modifications of the procedures may be usable for biotyping purposes. The adoption of specific microbial reference strains well-characterized in their susceptibility pattern could allow the control of reproducibility and reliability of the intra- and interlaboratory results by using a defined killer system.

4. Slight differences in the modalities of preparation of killer toxins neither affect the results obtained with the different sensitive microorganisms nor imply substantial variations in their carbohydrate or protein contents.

5. The (partially) purified killer toxin fractions proved to be stable when maintained at 4°C thus giving remarkable test reproducibility.

6. Concentrated crude killer toxins may be used in place of the (partially) purified fractions with similar results.

7. The standardized inoculum of sensitive yeast cells in a log phase is replaceable by a light suspension in sterile distilled water of the same yeast isolate without loss of sensitivity of the test.

8. The suspension of the microorganism to be tested may alternatively be inoculated by a swab onto the whole surface of the agar plates.

9. Alternatively, 50 µL drops of heavy distilled water suspension of each killer yeast grown on Sabouraud dextrose agar for

10. A conidial suspension of 0.5 McFarland diluted 1:10 in sterile distilled water may be used for *Aspergillus* sect. *Fumigati* strain differentiation *(40)*.

11. The same biotype defined by the killer system may be due either to the limited number of the killer yeasts or toxins used or to the same provenance of the investigated microorganisms.

12. In the computer-aided killer system, some checks must be preliminarily performed using reference control yeast isolates in order to verify the reliability of the groups obtained. As the comparison may reveal some variability, due either to errors in measuring the ID or to the susceptibility changes of yeast isolates to the toxin over a long period of time, such errors must be reduced by slightly changing the ranges in order to achieve a maximum error of 5% *(43)*. The storage of the data in the computer allows the rapid comparison of the result of any new isolate with those previously coded.

Also, 48 h at 25°C are placed on the surface of the agar containing the microbial isolate to be tested.

References

1. Odds, F.C. (1985) Biotyping of medically important fungi, in *Current Topics in Medical Mycology* (cGinnis, M.R., ed.), Springer-Verlag, New York-Berlin-Heidelberg, pp. 155–171.
2. Walker, J. (1950) Variation in *Microsporum canis* and *Microsporum audouinii*. *Br. J. Dermatol.* **62**, 395–401.
3. Hasenclever, H.F. and Mitchell, W.O. (1961) Antigenic studies of *Candida*. I. Observation of two antigenic groups in *Candida albicans*. *J. Bacteriol.* **82**, 570–573.
4. Ajello, L. and Cheng, S.-L. (1967) The perfect state of *Trichophyton mentagrophytes*. *Sabouraudia* **5**, 230–234.
5. McDonough, E.S. and Lewis, A.L. (1967) *Blastomyces dermatitidis*: production of the sexual stage. *Science* **156**, 528–529.
6. Wilson, D.E., Bennett, J.E., and Bailey, J.W. (1968) Serologic grouping of *Cryptococcus neoformans*. *Proc. Soc. Exp. Biol. Med.* **127**, 820–823.
7. Kwon Chung, K.J. (1972) Sexual stage of *Histoplasma capsulatum*. *Science* **175**, 326.
8. Hasegawa, A. and Usui, K. (1974) The perfect state of *Microsporum canis*. *Nippon Juigaku Zasshi* **36**, 447–449.
9. Marsik, F.J. and Parisi, J.T. (1971) Bacteriophage types and O antigen groups of *Escherichia coli* from urine. *Appl. Microbiol.* **22**, 26–31.
10. Smith, P.B. (1972) Bacteriophage typing of *Staphylococcus aureus*, in *The Staphylococci* (Cohen, J.O. ed.), Wiley Interscience, New York, pp. 431–441.
11. Lindberg, R.B. and Latta, R.L. (1974) Phage typing of *Pseudomonas aeruginosa*: clinical and epidemiologic considerations. *J. Infect. Dis.* **130**(Suppl), S33–S42.
12. Aber, R.C. and Mackel, D.C. (1981) Epidemiologic typing of nosocomial microorganisms. *Am. J. Med.* **70**, 899–905.
13. Janda, J.M. and Bottone, E.J. (1981) *Pseudomonas aeruginosa* enzyme profiling: predictor of potential invasiveness and use as an epidemiological tool. *J. Clin. Microbiol.* **14**, 55–60.
14. Bauernfeind, A., Petermuller, C., and Schneider, R. (1981) Bacteriocins as tools in analysis of nosocomial *Klebsiella pneumoniae* infections. *J. Clin. Microbiol.* **14**, 15–19.
15. Fyfe, J.A., Harris, G., and Govan, J.R. (1984) Revised pyocin typing method for *Pseudomonas aeruginosa*. *J. Clin. Microbiol.* **20**, 47–50.
16. Tagg, J.R. (1984) Production of bacteriocin-like inhibitors by group A streptococci

of nephritogenic M types. *J. Clin. Microbiol.* **19**, 884–887.

17. Polonelli, L., Conti, S., Gerloni, M., Magliani, W., Chezzi, C., and Morace, G. (1991) Interfaces of the yeast killer phenomenon. *Crit. Rev. Microbiol.* **18**, 47–87.

18. Bevan, E.A. and Makower, M. (1963) The physiological basis of the killer character in yeast, in *Genetics Today, XI Int. Congr. Gen.* (Geerts, S.J. ed.), Pergamon Press, Oxford, pp. 202–203.

19. Philliskirk, G. and Young, T.W. (1975) The occurrence of killer character in yeasts of various genera. *Antonie Van Leeuwenhoek* **41**, 147–151.

20. Wickner, R.B. (1976) Killer of *Saccharomyces cerevisiae*: a double-stranded ribonucleic acid plasmid. *Bacteriol. Rev.* **40**, 757–773.

21. Stumm, C., Hermans, J.M., Middelbeek, E.J., Croes, A.F., and de Vries, G.J. (1977) Killer-sensitive relationships in yeasts from natural habitats. *Antonie Van Leeuwenhoek* **43**, 125–128.

22. Young, T.W. and Yagiu, M. (1978) A comparison of the killer character in different yeasts and its classification. *Antonie Van Leeuwenhoek* **44**, 59–77.

23. Bussey, H., Saville, D., Hutchins, K., and Palfree, R.G. (1979) Binding of yeast killer toxin to a cell wall receptor on sensitive *Saccharomyces cerevisiae*. *J. Bacteriol.* **140**, 888–892.

24. Kandel, J.S. and Stern, T.A. (1979) Killer phenomenon in pathogenic yeast. *Antimicrob. Agents Chemother.* **15**, 568–571.

25. Middelbeek, E.J., Stumm, C., and Vogels, G.D. (1980) Effects of *Pichia kluyveri* killer toxin on sensitive cells. *Antonie Van Leeuwenhoek* **46**, 205–220.

26. Bussey, H. (1981) Physiology of killer factor in yeast. *Adv. Microb. Physiol.* **22**, 93–122.

27. Gunge, N., Tamaru, A., Ozawa, F., and Sakaguchi, K. (1981) Isolation and characterization of linear deoxyribonucleic acid plasmids from *Kluyveromyces lactis* and the plasmid-associated killer character. *J. Bacteriol.* **145**, 382–390.

28. Hutchins, K. and Bussey, H. (1983) Cell wall receptor for yeast killer toxin: involvement of (1-6)-beta-D-glucan. *J. Bacteriol.* **154**, 161–169.

29. Tipper, D.J. and Bostian, K.A. (1984) Double-stranded ribonucleic acid killer systems in yeasts. *Microbiol. Rev.* **48**, 125–156.

30. Starmer, W.T., Ganter, P.F., Aberdeen, V., Lachance, M.A., and Phaff, H.J. (1987) The ecological role of killer yeasts in natural communities of yeasts. *Can. J. Microbiol.* **33**, 783–796.

31 Magliani, W., Conti, S., Gerloni, M., Bertolotti, D., and Polonelli, L. (1997) Yeast killer systems. *Clin. Microbiol. Rev.* **10**, 369–400.

32. Guyard, C., Dehecq, E., Tissier, J.P., Polonelli, L., Dei Cas, E., Cailliez, J.C., and Menozzi, F.D. (2002) Involvement of [beta]-glucans in the wide-spectrum antimicrobial activity of *Williopsis saturnus* var. *mrakii* MUCL 41968 killer toxin. *Mol. Med.* **8**, 686–694.

33. Kurtzman, C.P. (1984) Synonomy of the yeast genera *Hansenula* and *Pichia* demonstrated through comparisons of deoxyribonucleic acid relatedness. *Antonie Van Leeuwenhoek* **50**, 209–217.

34. Polonelli, L. and Morace, G. (1986) Reevaluation of the yeast killer phenomenon. *J. Clin. Microbiol.* **24**, 866–869.

35. Polonelli, L., Archibusacci, C., Sestito, M., and Morace, G. (1983) Killer system: a simple method for differentiating *Candida albicans* strains. *J. Clin. Microbiol.* **17**, 774–780.

36. Morace, G., Archibusacci, C., Sestito, M., and Polonelli, L. (1984) Strain differentiation of pathogenic yeasts by the killer system. *Mycopathologia* **84**, 81–85.

37. Polonelli, L., Dettori, G., Cattel, C., and Morace, G. (1987) Biotyping of micelial fungus cultures by the killer system. *Eur. J. Epidemiol.* **3**, 237–242.

38. Morace, G., Dettori, G., Sanguinetti, M., Manzara, S., and Polonelli, L. (1988) Biotyping of aerobic actinomycetes by modified killer system. *Eur. J. Epidemiol.* **4**, 99–103.

39. Morace, G., Manzara, S., Dettori, G., Fanti, F., Conti, S., Campani, L., Polonelli, L., and Chezzi, C. (1989) Biotyping of bacterial isolates using the yeast killer system. *Eur. J. Epidemiol.* **5**, 303–310.

40. Polonelli, L., Conti, S., Campani, L., and Fanti, F. (1990) Biotyping of *Aspergillus fumigatus* and related taxa by the yeast killer system, in *Modern Concepts in Penicillium and Aspergillus Classification* (Samson, R.A. and Pitt, J.I. eds.), Plenum Press Publishing Corporation, New York, pp. 225–233.

41. Polonelli, L., Menozzi, M.G., Campani, L., Gerloni, M., Conti, S., Morace, G., and

Chezzi, C. (1992) Anaerobic yeast killer systems. *Eur. J. Epidemiol.* **8**, 471–476.

42. Mulè, G., Stea, G., Logrieco, A., Conti, S., Polonelli, L., and Bottalico, A. (1994) Biotipizzazione di specie di *Fusarium* tossigene mediante impiego di lieviti antagonistici. *Micologia Italiana* **2**, 68–75.

43. Polonelli, L., Castagnola, M., Rossetti, D.V., and Morace, G. (1985) Use of killer toxins for computer-aided differentiation of *Candida albicans* strains. *Mycopathologia* **91**, 175–179.

44. Polonelli, L., Conti, S., Campani, L., Morace, G., and Fanti, F. (1989) Yeast killer toxins and dimorphism. *J. Clin. Microbiol.* **27**, 1423–1425.

45. Merz, W.G. (1990) *Candida albicans* strain delineation. *Clin. Microbiol. Rev.* **3**, 321–334.

46. Hunter, P.R. (1991) A critical review of typing methods for *Candida albicans* and their applications. *Crit. Rev. Microbiol.* **17**, 417–434.

47. Polonelli, L. and Morace, G. (1988) Killer systems and pathogenic fungi. *Eur. J. Epidemiol.* **4**, 415–418.

48. Polonelli, L., Conti, S., Magliani, W., and Morace, G. (1989) Biotyping of pathogenic fungi by the killer system and with monoclonal antibodies. *Mycopathologia* **107**, 17–23.

49. Boekhout, T. and Scorzetti, G. (1997) Differential killer toxin sensitivity patterns of varieties of *Cryptococcus neoformans*. *J. Med. Vet. Mycol.* **35**, 147–149.

50. Boekhout, T., van Belkum, A., Leenders, A.C., Verbrugh, H.A., Mukamurangwa, P., Swinne, D., and Scheffers, W.A. (1997) Molecular typing of *Cryptococcus neoformans*: taxonomic and epidemiological aspects. *Int. J. Syst. Bacteriol.* **47**, 432–442.

51. Buzzini, P. and Martini, A. (2000) Differential growth inhibition as a tool to increase the discriminating power of killer toxin sensitivity in fingerprinting of yeasts. *FEMS Microbiol. Lett.* **193**, 31–36.

52. Buzzini, P. and Martini, A. (2000) Utilisation of differential killer toxin sensitivity patterns for fingerprinting and clustering yeast strains belonging to different genera. *Syst. Appl. Microbiol.* **23**, 450–457.

53. Buzzini, P. and Martini, A. (2001) Discrimination between *Candida albicans* and other pathogenic species of the genus *Candida* by their differential sensitivities to toxins of a panel of killer yeasts. *J. Clin. Microbiol.* **39**, 3362–3364.

54. Maia, M.L., Dos Santos, J.I., Viani, F.C., Larsson, C.E., Paula, C.R., and Gambale, W. (2001) Phenotypic characterization of *Microsporum canis* isolated from cats and dogs. *Mycoses* **44**, 480–486.

55. Buzzini, P., Berardinelli, S., Turchetti, B., Cardinali, G., and Martini, A. (2003) Fingerprinting of yeasts at the strain level by differential sensitivity responses to a panel of selected killer toxins. *Syst. Appl. Microbiol.* **26**, 466–470.

56. Huerfano, S., Cepero, M.C., and Castaneda, E. (2003) Caracterizacion fenotipica de aislamientos ambientales de *Cryptococcus neoformans*. [Phenotype characterization of environmental Cryptococcus neoformans isolates]. *Biomedica* **23**, 328–340.

57. Buzzini, P., Turchetti, B., and Martini, A. (2004) Assessment of discriminatory power of three different fingerprinting methods based on killer toxin sensitivity for the differentiation of *Saccharomyces cerevisiae* strains. *J. Appl. Microbiol.* **96**, 1194–1201.

58. Corte, L., Lattanzi, M., Buzzini, P., Bolano, A., Fatichenti, F., and Cardinali, G. (2005) Use of RAPD and killer toxin sensitivity in *Saccharomyces cerevisiae* strain typing. *J. Appl. Microbiol.* **99**, 609–617.

Chapter 12

DNA Fingerprinting *Candida* Species

Claude Pujol and David R. Soll

Abstract

It is sometimes necessary to assess the genetic relatedness of isolates to identify the origin of an infection. In addition, evidence is accumulating that drug resistance can be associated with strains from a particular clade and that strains can exhibit anatomical specificity. It may, therefore, be valuable in the near future to screen for strains with a propensity for drug resistance. While a number of methods exist for genetically fingerprinting the infectious fungi, only a few provide the necessary resolution not only for distinguishing whether strains are highly related or unrelated, but also for grouping a strain in a particular clade. Here, we provide the procedures for performing the two methods that have proven most effective in the past 5 years: Southern blot hybridization of restriction fragments with complex probes and multilocus sequence typing (MLST).

Key words: DNA fingerprinting, restriction fragment-length polymorphism (RFLP), complex probes, multilocus sequence typing (MLST).

1. Introduction

The capacity to assess genetic relatedness among strains not only provides us with insights into the population structure and mode of propagation of *Candida* species *(1)*, but also with information on the epidemiology of infections *(2, 3)*. The discoveries that flucytosine resistance is due to a single nucleotide change in *FUR1*, the gene that encodes uracil phosphoribosyltransferase in *C. albicans*, and that this allele is restricted almost exclusively to isolates of clade 1 (Group 1) *(4)*, suggest that genetic typing may eventually become a component of diagnosis. This idea is reinforced by the observation that isolates of different clades may have a propensity for colonizing different anatomical locations of a host *(5)*. Hence, potential drug resistance and colonization propensities may eventually be assessed by genetically fingerprinting fresh isolates from patients at risk.

Over the past 20 years, a number of technologies have been used to assess the level of strain relatedness. These methods have included multilocus enzyme electrophoresis (MLEE), restriction fragment-length polymorphism (RFLP), Southern blot hybridization of restriction fragments with repetitive or "complex" probes, random amplification of polymorphic DNA (RAPD), electrophoretic karyotyping, and a variety of sequencing methods, including multilocus sequence typing (MLST) *(4)*. There are a variety of reasons why some of these methods are not very effective, while others are highly effective. For instance while RFLP can provide a resolvable banding pattern for prokaryotes, the pattern it provides for eukaryotes is usually difficult to resolve due to increased genome size. Electrophoretic karyotyping is problematic because of pattern variability resulting from reagents, sample preparation and electrophoretic conditions, and the highly unpredictable rate of change that can be variable between strains *(6)*. Hence, it provides poor measure of genetic relatedness. Some methods, like MLEE, are highly reproducible and quite effective in assessing levels of genetic relatedness, but are apparently too complicated for many mycologists to perform. On the other hand, Southern blot hybridization of restriction fragments with complex probes provides a highly reproducible method for assessing genetic relatedness. The information obtained with this latter method can discriminate not only clades that have evolved over a long period of time but also rapid and recent changes within a strain due to microevolution. These attributes are also true for MLST. Advances in automated sequencing and the recent availability of full genome sequences have resulted in the rapid development of MLST as the new genetic fingerprinting method of choice. Given the overall effectiveness of both hybridization of restriction fragments with complex probes and MLST, we have focused on these two methods in the present chapter.

2. Materials

2.1. Cell Culture and Genomic DNA Purification

1. Yeast–peptone–dextrose (YPD) medium: 1% yeast extract (Becton, Dickinson and Company, Sparks, MD, USA), 2% peptone (Becton Dickinson), and 2% dextrose (Fisher Scientific, Fair Lawn, NJ). The solution is autoclaved and then stored at room temperature.

2. Sorbitol potassium phosphate (SPP) solution is prepared by dissolving 1 M sorbitol into a 50 mM potassium phosphate solution. The potassium phosphate solution is prepared by adding 40.1 mL of 1 M K_2HPO_4 and 9.9 mL of 1 M KH_2PO4 to 1 L of final volume. The pH of the potassium phosphate solution is adjusted to 7.5 by adding drops of 1 M K_2HPO_4 or KH_2PO4 as needed. Autoclave the SPP solution. Store at room temperature.

3. Zymolyase solution: 200 mg of Zymolyase 20T (Seikagaku America, Ijamsville, MD, USA) are dissolved in 1 mL H_2O. Store at −20°C.

4. 2-Mercaptoethanol (Sigma, St. Louis, MO, USA).

5. Lysis buffer: 250 mM Tris-HCl, pH 7.5, 50 mM ethylenediamine tetraacetic acid (EDTA), 5% (w/v) sodium dodecyl sulfate (SDS). Store at room temperature.

6. Proteinase K (Roche Diagnostics Corporation, Indianapolis, IN, USA) is dissolved at 10 mg/mL in H_2O. Store at −20°C.

7. Phenol/chloroform (1:1). Store at 4°C.

8. Potassium acetate (5 M).

9. 2-Propanol.

10. 70% Ethanol (v/v).

11. 1X TE: 10 mM Tris-HCl, pH 7.5, 1 mM EDTA.

12. Ribonuclease A (RNase) from Sigma. Prepare a solution of 10 mg of RNase A in 1 mL 10 mM Tris-HCl (pH 7.5), 15 mM NaCl. Store at −20°C.

13. Sodium acetate (3 M), pH 4.5.

14. 100% Ethanol.

2.2. Southern Blot Hybridization

1. Restriction enzyme. *Eco*RI is usually used, but other enzymes or combination of enzymes may be better suited in function of the species analyzed and the probe used (*see* **Note 1**).

2. 1X Tris Borate EDTA (TBE) buffer: prepare a 10X stock solution – 0.89 M Tris base, 0.89 M boric acid, and 10 mM EDTA; adjust pH to 8.0.

3. Agarose for gel electrophoresis (i.e., molecular biology grade with a low electroendosmosis).

4. 10X DNA loading dye solution: 0.1% (w/v) bromophenol blue, 0.1% (w/v) xylene cyanol, 15% (w/v) Ficoll 400, 10 mM Tris-HCl (pH 7.5), and 50 mM EDTA (pH 8.0).

5. Ethidium bromide solution (10 mg/mL).

6. 0.4 M HCl.

7. Denaturation buffer: 0.5 M NaOH, 1.5 M NaCl.
8. 25X SSC: 3.75 M NaCl, 0.37 M sodium citrate, pH 7.0.
9. Hybond N$^+$ nylon membrane (Amersham Biosciences, Piscataway, NJ).
10. 20X SPPE: 3 M NaCl, 0.2 M $NaH_2PO_4.H_2O$, 20 mM EDTA, pH 7.0.
11. Hybridization buffer: 5X SSPE, 5% (w/v) dextran sulfate (sodium salt, produced from dextran with an average molecular weight of 5000,000; Research Products International Corp., Mt. Prospect, IL, USA), 0.3% (w/v) SDS.
12. Calf thymus DNA (sodium salt; Sigma): 10 mg/mL in sterile H_2O. Shear the DNA by sonication to obtain an average fragment size of less than 2 kb.
13. Wash solution: 2X SSC, 0.2% (w/v) SDS.
14. A species-specific complex DNA fingerprinting probe (*see* **Note 2**).
15. Random primer labeling kit (e.g., DECAprime II kit from Ambion, Inc., Austin, TX, USA).
16. $[\alpha\text{-}^{32}P]dCTP$.
17. Sephadex G-50 column (5′–3′, Boulder, CO, USA).
18. Autoradiography film (e.g., Kodak BioMax XAR Film; Eastman Kodak Company, Rochester, NY, USA).

2.3. Multilocus Sequence Typing

2.3.1. Polymerase Chain Reaction Amplification

1. Taq DNA polymerase and its provided buffer (New England BioLabs, Inc., Ipswich, MA, USA).
2. Deoxynucleotide 5′-triphosphate (dNTP); 2 mM each of deoxyadenosine 5′-triphosphate (dATP), deoxythymidine 5′-triphosphate (dTTP), deoxycitidine 5′-triphosphate (dCTP), and deoxyguanosine 5′-triphosphate (dGTP) from Invitrogen (Carlsbad, CA, USA).
3. Locus specific oligonucleotide primer pairs (5 μM each).

2.3.2. Agarose Gel Analysis and Amplicon Purification

1. DNA molecular-weight markers (Promega, Madison, WI, USA).
2. 0.5X Tris Borate EDTA (TBE) buffer: 10X stock solution – 0.89 M Tris base, 0.89 M boric acid, and 10 mM EDTA; adjust pH to 8.0.
3. Agarose for gel electrophoresis (i.e., molecular biology grade with a low electroendosmosis).
4. 10X DNA loading dye solution 0.1% (w/v) bromophenol blue, 0.1% (w/v) xylene cyanol, 15% (w/v) Ficoll 400, 10 mM Tris-HCl (pH 7.5), and 50 mM EDTA (pH 8.0).

5. Ethidium bromide solution (10 mg/mL).

6. Commercial kit for direct purification and cleanup of DNA from polymerase chain reaction (PCR) products (e.g., QIAquick PCR Purification Kit from QIAGEN, Valencia, CA, USA), or commercial kit for extraction of DNA fragment and cleanup from agarose gel (e.g., QIAquick Gel Extraction Kit from QIAGEN).

2.3.3. DNA Sequencing

1. Sequencing primers (5 µM each).

2. Big Dye terminator ready reaction mix (e.g., version 3.1 from Applied Biosystems, Foster City, CA, USA). Dilute as follows: 1 µL Big Dye 3.1, 2 µL Big Dye buffer (comes with Big Dye), 1 µL distilled water, and 0.5 µL dimethylsulfoxide (DMSO from Sigma).

3. 95% ethanol (v/v).

4. Sodium acetate (3 M), pH 5.2.

5. 70% ethanol (v/v).

6. Formamide.

3. Methods

3.1. DNA Purification

The quality of the DNA used for fingerprinting in methods based on Southern blot hybridization with complex repetitive probes has a strong impact on the quality of the patterns obtained. The method described here, while not the fastest one, has been shown to be effective in obtaining large quantities of total DNA containing high-molecular-weight genomic DNA suitable for Southern blot DNA fingerprinting or PCR-based methods.

1. Grow yeast cells overnight in 3 mL YPD.

2. Harvest cells in a 1.5-mL microfuge tube by centrifugation at 15,000 × g for 1 min. Repeat once to harvest remaining cells.

3. Wash cells once by resuspending them in 0.8 mL of SPP and pelleting them at 15,000 × g for 1 min.

4. Resuspend cells in 0.8 mL of SPP. Add 1 µL of 2-mercaptoethanol and 4 µL of zymolyase solution per tube and incubate for 1 h at 37°C. Upon degradation of the yeast cell wall due to the zymolyase and 2-mercaptoethanol treatment, spheroplasts are obtained. Passed that stage, do not vortex.

5. Centrifuge spheroplasts at 15,000 × g for 1 min. Resuspend the spheroplast pellet in 0.8 mL of SPP (*see* **Note 3**). Repeat

this step. Add 0.5 mL of lysis buffer and 10 µL of proteinase K to the spheroplast pellet. Resuspend pellet (*see* **Note 3**) and incubate at 55°C for 2 h (or overnight).

6. Extract once with an equal volume of phenol–chloroform. Centrifuge for 5 min at 15,000 × *g* and transfer the supernatant into a new tube.

7. Add 200 µL of potassium acetate. Mix by inverting the tube four to six times. Incubate in ice for 15 min. Centrifuge at 15,000 × *g* for 10 min at 4°C. Transfer the supernatant into a new tube.

8. Repeat **Step 6**.

9. Add an equal volume of 2-propanol and precipitate the DNA by inverting the tube four to six times. Pellet the DNA at 15,000 × *g* for 30 s. Wash the pellet once with 70% ethanol.

10. Dry the DNA pellet (on the bench or with a speedvac) and resuspend in 400 µL of TE buffer containing 4 µL of RNase A. Incubate overnight at 37°C.

11. Repeat **Step 6**.

12. Add 40 µL of sodium acetate and 800 µL of 100% ethanol to the DNA solution. Precipitate the DNA by inverting the tube four to six times and pellet the DNA at 15,000 × *g* for 30 s. Wash the pellet once with 70% ethanol. Resuspend the pellet in 80 µL TE.

3.2. RFLP Analysis with a Species-Specific Complex DNA Fingerprinting Probe

RFLP DNA fingerprinting identified by hybridization with a species-specific complex DNA probe is a very effective method for analyzing genetic diversity at the subspecies level or microevolution *(2)*. Due to their species-specificity, these probes are also particularly useful to ascertain the species status of a strain. In this protocol, genomic DNA is cut with a restriction enzyme, separated by electrophoresis, transferred onto a nylon membrane, and hybridized with a specific DNA probe (*see* **Note 2**). The result is a complex banding pattern obtained only with the genomic DNA of strains from the species for which the DNA probe is specific. The DNA from strains of a distinct species will either not give any banding pattern *(7–11)*, or where the two species are closely related, give a very distinctive pattern containing only a few, often less intense, bands *(10, 11)*. In addition, banding patterns generated by using DNA probes are easily amenable to cluster analysis [*see* **Note 4** *(2)*]. This allows for a more detailed examination of the relationship between strains and groups of strains that is often needed when dealing with closely related species.

1. Prepare DNA following the instructions provided in **Section 3.1.**

2. Digest 3 µg of genomic DNA with 1 µL of *Eco*RI (*see* **Note 1**) in a final volume of 20 µL. Incubate overnight at 37°C. Add 2.5 µL of loading dye. In order to calibrate the migration of the bands between gels, load a reference strain in the two outer lanes.

3. Separate the digested fragments on a 1X TBE, 0.8% (w/v) agarose gel containing 200 µg/L of ethidium bromide (*see* **Note 5**). Run the gel at 60 V overnight. Stop the electrophoresis when the bromophenol blue has migrated close to 18 cm from the loading wells.

4. Depurinate the gel by placing it in a 0.4 M HCl solution and incubate under mild agitation for 30 min. This treatment will infrequently nick the DNA, reducing the size of the fragments and allowing for a better transfer from the gel to the membrane. Do not allow this treatment to proceed for more than 30 min.

5. Denature the DNA by incubating the gel in denaturation buffer for 30 min. A longer treatment will not affect the procedure.

6. Transfer the gel to a Hybond N⁺ membrane following the manufacturer's instructions with the following modification: use 12X SSC for the blotting buffer. Allow the transfer to proceed overnight. Fix the membrane by UV cross-linking using a transilluminator.

7. Prehybridize the membrane for 4 h in 30 mL of 65°C hybridization buffer containing 0.1 mg/mL of heat-denatured calf thymus DNA. Carry the prehybridization in a 65°C shacking water bath. Several membranes can be treated together. If these are performed in heat-sealable bags, up to four membranes can be treated together in a volume of buffer of 50–60 mL.

8. Prepare a ^{32}P-dCTP-labeled DNA probe by following the random primer labeling kit manufacturer's recommendations. Unincorporated radioactivity is removed by filtration on a sephadex G-50 column. Control labeling efficiency with a scintillation counter and add 5×10^6 cpm of heat denatured labeled probe to the hybridization buffer. Hybridize at 65°C for 4 h. A longer incubation time (e.g., overnight) will not affect results.

9. Discard the hybridization buffer and wash the membrane 3×10 min in 500 mL of wash buffer. Carry these steps in a 45°C shaking water bath. Up to four membranes can be washed together in the same volume of wash buffer.

10. Enclose the membrane in saran wrap to keep it from drying, and expose it to an autoradiography film. Exposure time will vary in function of the strength of the radioactive signal (3 h to 4 days).

11 Membranes can be stripped off the binding probe and reused with a different probe as long as they are not entirely dry. Follow the membrane manufacturer's instructions. Membranes can be stored at room temperature for several months.

3.3. Multilocus Sequence Typing

In recent years, automatic sequencing has become a fast and relatively inexpensive method. In addition, sequence data can be easily shared among the scientific community and are ideal to create valuable databases. For these reasons, the use of effective MLST-based approaches is rapidly expanding to the analysis of the genetic diversity of further species. While the MLST method has mostly been used to analyze subspecies genetic diversity, it is also useful to determine a strain species. In particular, MLST can be very effective when analyzing closely related species. Tavanti and colleagues *(12)* have recently separated *Candida parapsilosis* into three closely related species, *C. parapsilosis*, *C. orthopsilosis*, and *C. metapsilosis*, based on this technique. The method involves amplifying a DNA fragment by PCR and sequencing that locus. Two possible outcomes can arise when a strain is from a species distinct from the presumed one: (i) some or all loci will not be amplified by PCR due to mutations of the sequences recognized by the primers; (ii) the DNA sequence of the PCR products obtained will show a lower homology than the one observed between strains of the same species. For the latter outcome, sequence homologies of less than 90% over several loci may be indicative of different species. The proportion of loci falling into each category will vary in function of the relatedness of the species. The latter outcome will be more frequent as the two species are more closely related.

The MLST approach requires a previous knowledge of the gene sequences used for analysis. This step is hardly a hindrance anymore given the ongoing advances in genome sequences. For some species, MLST schemes have already being described and sequence databases implemented (available at http://www.mlst.net/databases/default.asp or http://pubmlst.org/). The use of loci previously analyzed by others is highly recommended as new datasets can easily be incorporated to existing databases and larger comparisons performed. However, for species where no MLST scheme has been implemented yet, several genes need to be screened in different isolates in order to find a set of about six loci presenting the level of polymorphism suitable for use with MLST. As a general rule, these genes should not include virulence factors or resistance determinants, as selective pressure may bias population genetics analyses (*see* **Note 6**).

The desired length of the amplified sequences to be analyzed should be between 500 and 650 bp. DNA fragments of this length can easily be synthesized by PCR and the full-length sequence obtained with a single sequencing run (*see* **Note 7**).

3.3.1. PCR Amplification of Loci

1. Prepare DNA following the instructions provided in **Section 3.1.**
2. PCRs are performed in a final volume of 25 µL containing 2.5 ng of template DNA, 1 U of *Taq* DNA polymerase, 2.5 µL of 10X PCR buffer, 0.2 mM of each dNTP, and 0.2 µM of each primer.
3. The PCR reaction conditions are as follows: 7 min at 94°C, 30 cycles of 1 min at 94°C, 1 min at an annealing temperature relevant for the primers used (this may vary from 50°C to 60°C in function of the T_m of the primers used), and 1 min at 74°C, followed by 10 min at 74°C.

3.3.2. Analysis and Purification of the Amplified Fragments

1. Use 5 µL of the PCR reaction mixed with 4 µL of distilled water and 1 µL of loading dye to analyze the purity and size of the fragments obtained. Fragments are separated in a 0.8% (w/v) agarose 0.5X TBE gel containing 200 µg/L of ethidium bromide. Run molecular-weight markers alongside to estimate the size of the fragments.
2. When the fragments obtained are pure (the PCR reaction does not contain extra bands), they can be purified directly by using a direct purification and cleanup kit. Otherwise, the expected bands can be extracted from a preparative agarose gel by using a kit for extraction of DNA fragment and cleanup from agarose gel.

3.3.3. DNA Sequencing

The primers used for the PCR reactions can be used for sequencing if the solution containing the amplified fragment does not include any nonspecific fragments. Otherwise primers hybridizing inside the amplified fragment of interest can be used (nested primers). Sequence all loci in both directions. This is particularly critical when analyzing loci from diploid organisms (*see* **Note 8**).

1. Prepare a reaction mix for 10 µL final volume containing 4.5 µL Big Dye mix, 15 ng PCR template (as a general rule use 2.5 ng/100 bp of amplified product), and 6 pmol of primer.
2. Use the following thermal cycling conditions: 4 min at 96°C, followed by 30 cycles of 10 s at 96°C, 5 s at 50°C, and 2 min at 60°C. Ramp to 4°C and hold.
3. To the 10 µL sequencing reaction, add 26 µL of distilled H_2O, 64 µL nondenatured 95% ethanol, and 3 µL of 3 M sodium acetate. The final ethanol concentration should be 60±3%. Centrifuge at 3000 × *g* for 30 min.

4. Wash the pellet with 100 μL of 70% ethanol and dry pellet. Resuspend in 10 μL formamide.

The reaction products can then be analyzed using an automated DNA sequencer according to the manufacturer's instructions (*see* **Note 9**).

4. Notes

1. With many complex DNA fingerprinting probes developed to date, the restriction enzyme *Eco*RI has been shown to provide patterns with the required complexity *(7–10)*. Nevertheless, in function of the probe and the species analyzed, the patterns obtained with this enzyme may not always present the finest complexity to answer the question asked in a given study. Other enzymes or combinations of enzymes can be assessed to increase pattern complexity. For example, in *C. parapsilosis* a combination of *Eco*RI and *Sal*I was shown to be more effective *(11)*.

2. Species-specific probes are available for a number of *Candida* and *Aspergillus* species *(7–11)*. When no complex species-specific DNA probe is available for the species of interest, a probe must be isolated, characterized, and verified to provide an accurate measure of genetic relationship *(2)*. A protocol for the isolation of species-specific DNA fingerprinting probes can be found in Lockhart et al. *(13)*.

3. To resuspend spheroplast pellets, use a 1-mL micropipette. When resuspending spheroplasts in SPP, draw the pellet into the 1 mL tip back-and-forth until the solution appears homogeneous. When spheroplasts are in the lysis buffer, homogeneity cannot be achieved because the spheroplasts start lysing and the pellet becomes viscous. Nevertheless, it is better to loosen the pellet by drawing it into a cut tip once or twice.

4. Dendrograms based on banding patterns generated by Southern blot hybridization of moderately repetitive sequences have been shown to provide a particularly useful representation of the relationship observed between strains and groups of strains *(2, 14)*. The similarity coefficient most commonly used (S_{AB}) is based on band position:

$$S_{AB} = 2E/(2E + a + b),$$

where E is the number of bands shared between strains A and B, a is the number of bands unique to strain A, and b is the number of bands unique to strain B. An S_{AB} of 1.00 represent identical banding patterns, an S_{AB} of 0.00 represents patterns with no matching bands, and S_{AB}s of 0.01–0.99 represents banding patterns of increasing similarity.

S_{AB}s can be computed between every pair of isolates, a matrix of S_{AB}s generated, and a dendrogram created based on that matrix. Many algorithms can be applied. The Unweighted Pair Group Method with arithmetic Average (UPGMA) is one of the most commonly used (15).

5. The ideal concentration of agarose may vary as a function of the fragment sizes expected in order to achieve optimum separation. For an example, in *C. glabrata* were most of the fragments sizes obtained by using the probes Cg6 or Cg12 were in the high–molecular-weight range (8–20 kb), 0.65% agarose gels were used to obtain better results (9).

6. The MLST methodology was originally developed to analyze bacterial populations. Due to the high number of polymorphisms found in these organisms, the loci used have been restricted to coding regions (16). In fungal species, the number of polymorphic nucleotide positions is usually considerably lower and the use of more polymorphic noncoding sequences may be of particular interest when analyzing populations with a low genetic diversity. The analysis of hypervariable loci can also be used to assess microevolution.

7. To reduce the probability of nonspecific annealing and spurious PCR products, high melting temperatures ($51°C \leq T_m \leq 61°C$) are recommended for both forward and reverse primers. Higher annealing temperatures can be used with elevated T_m. When conserved regions of the loci analyzed are known, primers should be designed to hybridize to those regions encompassing variable sequences.

8. The MLST method has primarily being used to analyze haploid microorganisms where the sequences generated identify unambiguous alleles. While MLST can also be used to analyze populations of diploid microorganisms [e.g., *C. albicans* (17)], the interpretation of the results presents unique challenges. In diploid microorganisms, double peaks in the sequencing traces will be indicative of heterozygous nucleotide positions. These double peaks are most often easily identifiable. Nevertheless, in certain cases, one of the two peaks may be misinterpreted as background signal. This can be due to a differential amplification of the two alleles during PCR, or because one of the nucleotides gives a stronger sequencing signal. To limit this source of error, sequencing traces obtained in both directions need to be carefully examined. When doubt persists, it is better to PCR the locus again. By sequencing the PCR products directly, the identification of individual alleles is not possible when more than one nucleotide position is heterozygous. To ascertain the two alleles, cloning and sequencing of at least one of them is required. The second allele can then be inferred by deduction. While this strategy is relatively straightforward

when heterozygosities are due to single nucleotide polymorphisms (SNPs), the analysis of loci containing insertion/deletion polymorphisms will be much more problematic, and such loci should be avoided with diploid microorganisms.

9. A number of free softwares are available for the analysis of MLST data, including START 2 (http://pubmlst.org/software/analysis/start2/, to perform data summaries, lineage assignments, tests for recombination, and tests for selection) and PHYLIP 3.65 (http://evolution.genetics.washington.edu/phylip.html, for phylogenetic analyses).

Acknowledgments

The work from the Soll lab was supported by NIH grants AT2392 and DEO4219.

References

1. Pujol, C., Dodgson, A., and Soll, D. R. (2005) Population genetics of ascomycetes pathogenic to humans and animals. In: Evolutionary Genetics of Fungi (Xu, J.-P., ed.), Horizon Scientific Press, Norfolk, UK, pp. 149–188.
2. Soll, D. R. (2000) The ins and outs of DNA fingerprinting the infectious fungi. Clin. Microbiol. Rev. 13, 332–370.
3. Soll, D. R., Lockhart, S. R., and Pujol, C. (2007) Laboratory procedures for the epidemiological analysis of microorganisms. In: Manual of Clinical Microbiology 9th Edition (Murray, P. R., Baron, E. J., Pfaller, M. A., Jorgensen, J. H., Landry, M. L., eds.), ASM Press, Washington, D.C. pp. 129–151.
4. Dodgson, A. R., Dodgson, K .J., Pujol, C., Pfaller, M. A., and Soll, D. R. (2004) Clade-specific flucytosine resistance is due to a single nucleotide change in the FUR1 gene of Candida albicans. Antimicrob. Agents Chemother. 48, 2223–2227.
5. Tavanti, A., Davidson, A. D., Fordyce, M. J., Gow, N. A. R., Maiden, M. C. J., and Odds, F. C. (2005) Population structure and properties of Candida albicans, as determined by multilocus sequence typing. J. Clin. Microbiol. 43, 5601–5613.
6. Ramsey, H., Morrow, B., and Soll, D. R. (1994) An increase in switching frequency correlates with an increase in recombination of the ribosomal chromosomes of Candida albicans strain 3153A. Microbiology 140, 1525–1531.
7. Girardin, H., Latgé, J., Srikantha, T., Morrow, B., and Soll, D. R. (1993) Development of DNA probes for fingerprinting Aspergillus fumigatus. J. Clin. Microbiol. 31, 1547–1554.
8. Joly, S., Pujol, C., Schröppel, K., and Soll, D. R. (1996) Development of two species-specific fingerprinting probes for broad computer-assisted epidemiological studies of Candida tropicalis. J. Clin. Microbiol. 34, 3063–3071.
9. Lockhart, S. R., Joly, S., Pujol, C., Sobel, J., Pfaller, M. A., and Soll, D. R. (1997) Development and verification of fingerprinting probes for Candida glabrata. Microbiology 243, 3733–3746.
10. Joly, S., Pujol, S., Rysz, M., Vargas, K., and Soll, D. R. (1999) Development and characterization of complex DNA fingerprinting probes for the infectious yeast Candida dubliniensis. J. Clin. Microbiol. 37, 1035–1044.
11. Enger, L., Joly, S., Pujol, C., Simonson, P., Pfaller, M., and Soll, D. R. (2001) Cloning and characterization of a complex DNA fingerprinting probe for Candida parapsilosis. J. Clin. Microbiol. 39, 658–669.
12. Tavanti, A., Davidson, A. D., Gow, N. A. R., Maiden, M. C. J., and Odds, F. C. (2005)

Candida orthopsilosis and *Candida metapsilosis* spp. nov. to replace *Candida parapsilosis* Groups II and III. *J. Clin. Microbiol.* **43**, 284–292.

13. Lockhart, S. R., Pujol, C., Joly, S., and Soll, D. R. (2001) Development and use of complex probes for DNA fingerprinting the infectious fungi. *Med. Mycol.* **39**, 1–8.

14. Pujol, C., Joly, S., Lockhart, S. R., Nöel, S., Tibayrenc, M., and Soll, D. R. (1997) Parity among the randomly amplified polymorphic DNA method, multilocus enzyme electrophoresis, and Southern blot hybridization with the moderately repetitive DNA probe Ca3 for fingerprinting *Candida albicans*. *J. Clin. Microbiol.* **35**, 2348–2358.

15. Sneath, P. and Sokal, R. (1973) The Principles and Practice of Numerical Classification. W. H. Freeman & Co., San Francisco.

16. Maiden, M. C., Bygraves, J. A., Feil, E., Morelli, G., Russell, J. E., Urwin, R., Zhang, Q., Zhou, J., Zurth, K., Caugant, D. A., Feavers, I. M., Achtman, M., and Spratt, B. G. (1998) Multilocus sequence typing: a portable approach to the identification of clones within populations of pathogenic microorganisms. *Proc. Natl. Acad. Sci. USA* **95**, 3140–3145.

17. Bougnoux, M. E., Morand S., and d'Enfert, C. (2002) Usefulness of multilocus sequence typing for characterization of clinical isolates of *Candida albicans*. *J. Clin. Microbiol.* **40**, 1290–1297.

Part V
Genomics and Proteomics

Chapter 13

The Application of Tandem-Affinity Purification to *Candida albicans*

Chris Blackwell and Jeremy D. Brown

Abstract

Tandem-affinity purification (TAP) tagging systems, developed by the research group of Bertrand Seraphin and others, are a means of isolating physiologically relevant protein and protein–nucleic acid complexes. Where complete (or nearly complete) genome sequence data are available for the organism from which the complexes are isolated, their components can be readily identified using mass spectrometry. The most widely used TAP-tag consists of a proximal calmodulin-binding peptide (CBP) and a distal repeated protein A IgG-binding domain with a cleavage site for the tobacco etch virus (TEV) protease positioned between this and the CBP. This tag is expressed as a co-translational fusion to the protein of interest. Purification is achieved under mild conditions through sequential affinity chromatography on IgG (eluting by proteolytic cleavage with TEV protease) and calmodulin (eluting by removal of Ca^{2+} ions required for the interaction) resins. The approach has been hugely successful for categorizing the interactome of *Saccharomyces cerevisiae*. Here, we present vectors for carrying out TAP-tagging in *Candida albicans* and a protocol for purification of complexes containing TAP-tagged proteins.

Key Words: TAP-tag, purification, complex, interaction, mass spectrometry.

1. Introduction

1.1. Use of Protein "Tags" in *Candida albicans*

Despite the medical importance of *Candida albicans*, research into its cell biology has been somewhat hampered by lack of genetic tools and plasmids designed specifically to allow epitope tagging of proteins, analysis of promoter activity and purification of complexes of interacting proteins. Many of these limitations are now being overcome by a rapid rise in the number of such available reagents [e.g. *(1–11)*]. With a few exceptions *(6, 9)* a notable absence from published literature has been reagents and protocols for purification of native complexes. Here, we discuss the TAP-tagging and purification procedure for isolation of complexes and its application to *C. albicans*.

Defining the interactions of individual proteins and the components of multi-subunit complexes are key methods in modern cell biology. These approaches identify factors that collaborate, not only in three dimensions within static complexes, but also over time during the assembly, maturation and function of complex structures such as the spliceosome, ribosome and cytoskeleton. Other applications include the documentation of changes in interaction partners of proteins during response to changes in environmental conditions. Previously unsuspected links between different processes may be found, as can dual uses of proteins in separate cellular functions and pathways. Key methods that have allowed massive inroads into these areas have been genome-wide two-hybrid analysis *(12, 13)* and systematic affinity-capture coupled with mass spectrometry to identify interacting partners and components of complexes *(14, 15)*.

Efficient purification of native protein complexes under mild conditions has been achieved in several ways, and the most successful and widely used techniques involve the use of dual or tandem affinity tags. This provides the possibility of sequential mild purifications using different types of interaction with affinity matrices, thereby allowing highly specific recovery of the desired complexes. TAP-tagging procedures were initially designed for the isolation of native complexes in *Saccharomyces cerevisiae*, where they have been used exhaustively (one could almost say exhaustingly) to map the "interactome" of the organism *(16, 17)*. The approach has also been adapted successfully to higher eukaryotic cells, for example, in Ref. *(18)*.

In comparison, TAP-tagging has not been used widely with *C. albicans*. Only a few isolated analyses of the interactions of particular proteins have been reported *(6, 9)* and no large-scale experiments have been published. While there will be considerable similarity between complexes and processes in *S. cerevisiae* and *C. albicans*, detailed analysis of *C. albicans* will undoubtedly yield important information on the physiology of the organism. It is highly likely that should such a systematic approach be applied to *C. albicans*, novel information on pathways and complexes important for pathogenesis will be forthcoming. In addition, the information gleaned would provide a powerful evolutionary comparison to *S. cerevisiae*, as has been the case when transcript profiles of these organisms have been contrasted *(19, 20)*.

While molecular biological research into *C. albicans* suffers from the imposition of a diploid genome and a limited number of strains with multiple auxotrophic markers, the procedures used for TAP-tagging and purification of native complexes from *S. cerevisiae* are largely applicable to *C. albicans*. Amongst the published tags, those developed by Seraphin and co-workers have been adopted most widely. This is perhaps the easiest to use in practice, without the need for (relatively expensive) peptide competition to elute complexes bound to monoclonal antibodies used in a number of alternative systems. A limitation of the system is that the CBP–calmodulin interaction is inefficient in lipid phase. However, given the reduced complexity of protein mixtures in solubilized membrane preparations, sufficient purification for examination of complexes can often be achieved using the cleavable protein A tag in a single-step purification.

1.1.1. Adding a TAP-Tag to a Protein of Interest

We generated a series of integrative plasmids, primarily for use in *C. albicans*, derived from the plasmid CIpACT-CYC *(21)*, which contains the selectable marker *URA3*. These permit addition of either a cleavable protein A-tag *(7)* or the complete TAP-tag (**Fig. 13.1A**) to a protein of interest. The coding sequences used for these tags contained a single CTG codon (encoding leucine in the standard genetic code, but serine in *C. albicans*) within the tobacco etch virus (TEV) protease recognition sequence (ENLYFQG). This was changed to CTA to maintain the correct amino acid sequence. The protein A-tagging vectors are described in detail elsewhere *(7)*, and we will focus on TAP-tagging vectors here. Of these, two (CIpACT-N-TAP and CIpACT-C-TAP) allow insertion of an extra copy of the gene of interest along with an N- or C-terminal TAP-tag into the duplicated *RPS10* locus. All necessary promoter and terminator sequences to drive transcription are provided by the plasmid and expression of the resultant fusion is driven by the constitutive *ACT1* promoter *(22)*. Integration of *URA3* at *RPS10* restores wild-type virulence phenotypes to *ura3 C. albicans* strains *(23)*. A third plasmid, CIp-C-TAP, is appropriate for C-terminal tagging of one copy of the endogenous gene, leaving it under the control of its native promoter – and thereby ensuring correct expression levels. This can avoid artifactual interactions that are sometimes detected when proteins are over-expressed. In each situation, generation of a strain in which the tagged protein is the only one in the cell (by deletion of remaining endogenous copies of the gene) is extremely helpful as this: (a) reveals whether or not the tag inactivates or otherwise affects the function of the protein and (b) removes competition with the untagged protein for assembly into complexes.

In practice, expression of TAP-tagged proteins from integrated CIpACT-N-TAP and CIpACT-C-TAP derivatives is accomplished by fusing the open reading frame (ORF) encoding the protein of interest into the polylinker of the plasmid to generate an in-frame fusion with the sequences including the tag. Transformations are then performed *(7)* using plasmids linearised at one of the restriction sites within the *RPS10* fragment (usually

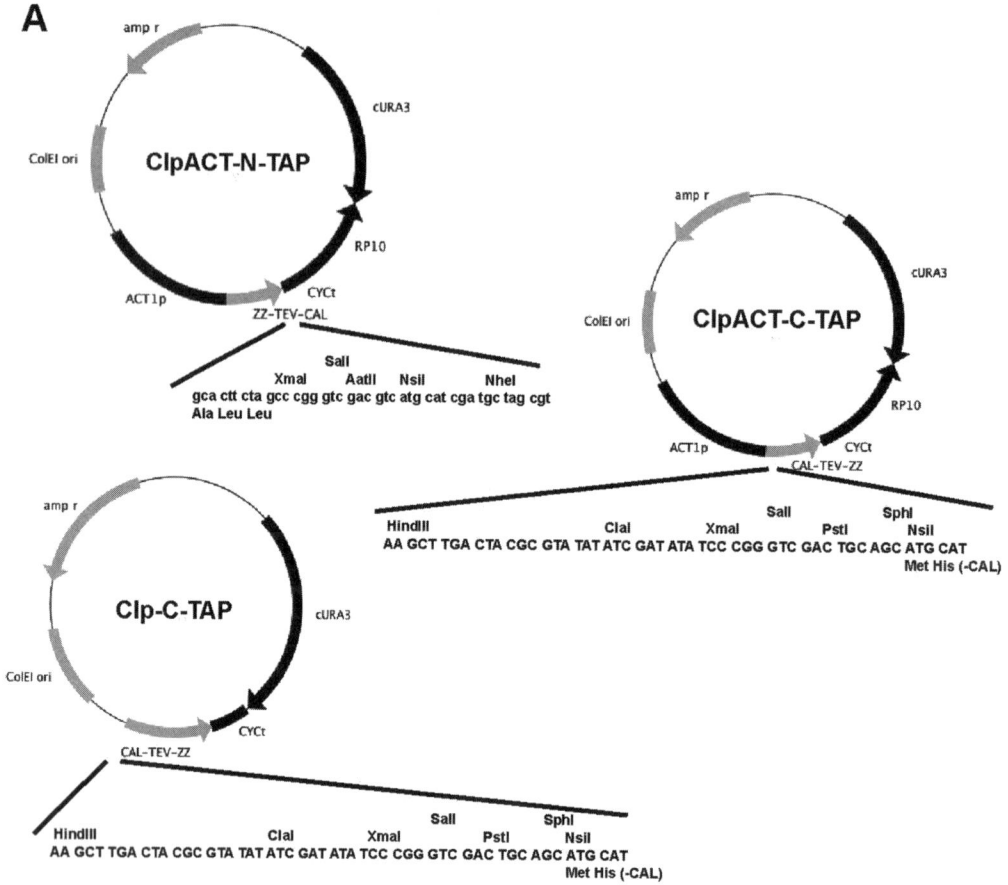

Fig. 13.1 TAP-tagging vectors. (**A**) Three vectors are shown. ClpACT-N-TAP and ClpACT-C-TAP, direct fusion of the TAP-tag to the 5' and 3' end of a gene inserted into the plasmid respectively and integration into the *RPS10* locus, where the gene is expressed from the *ACT1* promoter. Clp-C-TAP allows integration of sequences encoding the TAP-tag at the genomic locus of the gene of interest (see text for details). (**B**) Structure of PCR product amplified from Clp-C-TAP using oligonucleotides incorporating (a) 80 3' terminal nucleotides of the ORF encoding the protein to be tagged minus the stop codon and then TCC CGG GTC GAC TGC AGC, providing a linker SerArgValAspCysSer between the C-terminus of the synthesized protein and the tag and (b) the reverse complement of around 80 nucleotides downstream to the stop codon and the sequence CTACATATATACAAATCTAATAAAGTG, nucleotides 854-70 of the plasmid. The amplified product then contains a functional *URA3* gene as well as the TAG sequences in frame with the 3'end of the ORF. Maps were generated using WebPlasmid, version7.0.

*Stu*I or *Nco*I). For addition of the TAP-tag to a copy of the endogenous gene, we recommend inserting a fragment corresponding to the last 300-500 nucleotides of the target ORF (without the stop codon and in frame with the sequences encoding the tag) into

CIp-C-TAP. This provides sufficient homology to promote efficient site-specific integration of the plasmid into the genome. Ideally, the resulting plasmids are linearized using a unique restriction endonuclease site within the cloned fragment. Should there not be a unique restriction site in the coding sequence of the gene of interest, then linearizing the plasmid at a unique site in the polylinker at the 5' side of the cloned gene fragment will direct integration to the intended target site. As an alternative to cloning a fragment of the gene of interest into CIp-C-TAP, the tag and *URA3* gene can be amplified by polymerase chain reaction (PCR) (**Fig. 13.1B**). In all cases, transformants are selected by the presence of the *URA3* marker and can be conveniently screened for expression of the tagged protein by Western blotting. Many IgG bind protein A, and thus detection of the fusion protein is conveniently achieved using a peroxidase-coupled antibody (e.g., whole anti-rabbit IgG, Sigma) and enhanced chemiluminescence reagents (GE Healthcare). In our experience, we obtain a considerably higher level of correct, homologous integrations when plasmids that include gene fragments are transformed rather than PCR products.

1.1.2. Purification of TAP-Tagged Proteins

The protocol described below, adapted in part from *(24)*, has been used successfully in our laboratory and developed with parameters mentioned above in mind. For a number of the purifications that we have carried out, it has been important to remove ribosomes, or at least add sufficient salt – up to 500 mM, to reduce interactions with ribosomes. However, it is presented here without a ribosome-clearing spin (ribosomes do not inherently bind to IgG sepharose) and a "physiological" salt concentration of 150 mM under which most complexes are generally preserved.

2. Materials

2.1. Cell Growth and Harvesting

1. Yeast-extract–peptone–dextrose (YEPD) media: 1% (w/v) each of yeast extract and dextrose plus 2% (w/v) peptone.

2.2. Cell Lysis and Preparation of Extract

1. Lysis buffer: 150 mM KOAc, pH 7.4 (stock, usually 4 M is adjusted with KOH), 20 mM HEPES.KOH pH 7.4, 2 mM MgOAc, 1 mM EDTA, pH 8.0, 1 mM EGTA, pH 8.0, 0.02% Nikkol (octaethylene glycol monododecyl ether; Nikko Chemical Co., Tokyo, Japan), 1 mM PMSF, 2 mM dithiothreitol, mini EDTA-free protease inhibitor cocktail tablets (Roche) at a ratio of one tablet per 50 mL. Protease inhibitors and dithiothreitol are added just prior to use.

2. Waring model 8011 series blender with stainless steel chamber and vented steel lid, rather than the standard plastic lid, which is not resistant to liquid nitrogen temperatures.

2.3. Purification

1. IgG Sepharose™ 6 Fast Flow beads (GE Healthcare, Catalogue number 17-0969-01).
2. 0.8 × 4 cm Poly-Prep columns (Biorad 731-1550).
3. TEV buffer: as lysis buffer but without EGTA or protease inhibitor tablets.
4. AcTEV™ recombinant TEV protease (Invitrogen 12575 – 015/023).
5. CBB buffer: as lysis buffer but without EGTA, EDTA or protease inhibitor tablets and containing 2 mM $CaCl_2$.
6. Calmodulin affinity resin (Stratagene 214303-52).
7. CEB buffer: as lysis buffer (**Section 2.2**) but without protease inhibitor tablets and containing 20 mM EGTA.
8. 40% w/v Acrylamide and 2% w/v *bis*-acrylamide (available from a variety of suppliers, we use GE Healthcare), 1.5 M Tris.HCl, pH 8.8 (resolving gel buffer), 1.0 M Tris.HCl, pH6.8 (stacking gel buffer), 10% w/v SDS, N,N,N',N'-tetramethylethylenediamine (TEMED), 10% w/v ammonium persulphate (APS), which should be prepared fresh weekly, water saturated isobutanol (prepared by thoroughly mixing equal volumes of water and isobutanol; this will separate into two layers – use the upper layer). 10X Running buffer (250 mM Tris, 2.5 M glycine, 1% w/v SDS and the pH should be 8.3). All gel mixes must be handled with care, as acrylamide is a neurotoxin. Nitrile gloves are recommended.
9. SDS-PAGE (Laemmi) sample buffer: 2% SDS, 10% glycerol, 0.002% bromophenol blue and 62.5 mM Tris.HCl pH 6.8, 100 mM DTT (added just before use).
10. Colloidal Blue Staining Kit (Invitrogen LC6025).
11. SilverQuest™ Kit (Invitrogen LC6070).

3. Methods

The topic of this chapter is isolation of complexes of proteins, and thus we have focussed the methods from the point at which the desired strain has been generated. While many TAP-purifications are presented where cells from 2 to 3 L of culture are used, we find that unless multiple samples are being prepared, larger volumes are helpful as the increase in material can aid protein identification at later stages. An important control to include when setting up the method in the laboratory is a "mock" purification using a non-modified strain that does not express a TAP-tagged protein. While the high degree of specificity of the two affinity methods employed in the purification should ensure that the non-specific background is very low, it is important to have this information and images of gels from a mock purification to compare with experimental samples.

3.1. Cell Growth and Harvesting

1. Routinely we pool cells from 9 l cultures grown overnight with shaking at 30°C, in YPD (OD_{600} up to 10.0). Cells are harvested as quickly as possible by centrifugation at $3000 \times g$ (e.g. 5000 rpm in a Beckmann JA10 rotor) for 5 min, washed once with water and once with lysis buffer, before the resulting cell pellet is snap frozen in liquid nitrogen.

2. The preferred method of achieving rapid freezing is by forcing the cell pellet through a 50-mL syringe into liquid nitrogen contained in a Dewar flask. A small amount of lysis buffer, equal to about 5% of the total pellet volume, is mixed with the pellet before extrusion to facilitate the procedure. Ideally the effect is to generate "yeast spaghetti" which can be separated from the liquid nitrogen and stored at –80°C until required. Processing harvested cells in this manner greatly increases the speed at which cells can be ground or thawed during the initial stages of cell lysis.

3.2. Cell Lysis and Preparation of Extract

There are many methods to break yeast cells. Physical lysis is preferred to enzymatic methods, as agents used for cell wall digestion, such as zymolyase, have been reported to remove TAP tags from target proteins (24), in addition to being relatively inefficient at degrading the cell wall of *C. albicans*. We have successfully used two methods – "bead beating" (*see* **Note 1**) and "blending." Blending allows for quicker processing and is thus our preferred method. Whichever method of lysis is chosen, it is important to avoid introducing operator borne contaminants such as keratin, ribonucleases, proteases, etc. at any stage of the process. It is, therefore, extremely important to wear suitable gloves (e.g., Nitrile) during all stages of the purification and when handling equipment/tubes that will come into contact with experimental samples.

3.2.1. Blending

1. Up to ~150 g of frozen "yeast spaghetti" can be processed at one time in the blender, and the cells are kept cold by the regular addition of small amounts of liquid nitrogen to the chamber containing the frozen cells. Lysis is achieved by delivering 20 low-speed pulses, each of between 30 and 40 s, to the cells while maintaining pressure on the lid of the container to

prevent material escaping. The procedure is performed in the fume hood, and the apparatus and work area is thoroughly cleaned after use.

2. Following blending, the resultant fine powdered material is emptied into a sterile beaker placed in a shallow 37°C water bath.

3. As soon as the blended cells start to thaw, lysis buffer (at room temperature) is added in the ratio of 1 mL/g of starting material (*see* **Note 2**). It is important that the buffer contains protease inhibitors at this stage and that they, along with DTT, are added just before the buffer is used.

4. Once the buffer is added, the beaker is placed on ice and the preparation swirled gently until all lumps of frozen material are completely thawed. The lysate is then divided into 50 mL polycarbonate tubes and centrifuged at $25,000 \times g$ (e.g., 15,000 rpm in a Beckmann JA20 rotor) for 15 min at 4°C. The supernatant from this spin forms the "total cell extract."

3.2.2. Ultracentrifugation

1. It is a common practice to ultra-centrifuge total cell extracts before proceeding with the purification protocol. This removes cell debris, aggregated material and the lipidic phase from the total cell extract. Typically this is $100,000 \times g$ (e.g., 36,000 rpm in a Beckmann Type Ti45 rotor) for 60 min at 4°C. Upon completion of any ultra-centrifugation step, the clear middle portion of each sample is carefully pipetted out, and transferred to pre-chilled sterile 50 mL polypropylene screw-top tubes, being sure to avoid either the cloudy lipidic phase at the top or the pelleted material at the bottom of each tube.

3.3. Purification

3.3.1. Purification on IgG-Sepharose

1. The resulting clarified extract collected after the centrifugation step is incubated with IgG-sepharose beads pre-equilibrated in lysis buffer. Usually 0.5 mL of 50/50 beads/buffer slurry is added to each 50 mL of extract. However, the total volume of packed resin should not exceed 2 mL for a single preparation at the scale used here.

2. Incubation of the extract with the IgG beads is carried out with gentle agitation (e.g., on a continuous roller) at 4°C for at least 2 h, and may be performed overnight (*see* **Note 3**).

3. The beads are recovered from these incubations by centrifugation of the tubes at $1000 \times g$ for 2 min. The majority of the extract is carefully removed by pipetting, to leave the beads and about 5 mL of extract in each tube. The remaining extract, together with all the beads from each tube, is transferred to a column. As the extract drains from the column under gravity,

the beads pack in the column. The resulting column is then washed with 50 mL of lysis buffer (at the same salt concentration as the extract) and then 50 mL of TEV cleavage buffer.

4. After this, the base of the tube is then blocked. AcTEV™ protease of 200 units is added to 1 mL of TEV cleavage buffer, which is then pipetted directly onto the column and mixed with the beads. The column is then incubated at room temperature for 60 min (*see* **Note 4**).

5. Following this incubation, the TEV eluate is collected by unblocking the column and adding a further 4 mL of TEV cleavage buffer, yielding a 5 mL eluate in total. An appropriate amount (250–500 µL) is retained for analysis.

3.3.2. Purification on Calmodulin Affinity Resin

1. Calcium chloride is assed to the TEV eluate to 3 mM (from a 1 M stock), and the total volume is increased to 8 mL by the addition of calmodulin-binding buffer (CBB). This sample is then added to 0.5 mL of calmodulin affinity resin pre-equilibrated in CBB, in a 15-mL polypropylene screw-top tube. The sample is incubated with gentle agitation for 60 min at 4°C, before being transferred to a fresh 0.8 × 4 cm poly-prep column.

2. Once the sample has drained through the column, the beads are washed with 30 mL CBB, before the bound material is eluted in six separate 0.5 mL aliquots of calmodulin elution buffer (CEB). The peak of purified material is usually in fractions of 2 and 3.

3.4. Analysis of Purified Material

The end-goal of most purifications through the TAP-procedure is identification of factors co-purifying with the tagged protein and hence putative interacting partners. For this, most laboratories carry out SDS-PAGE gel electrophoresis of the purified material, stain the gel to identify candidate bands, digest these in-gel with trypsin and perform MALDI-TOF mass spectrometry of the digested material. Proteins are then identified by comparing the experimental masses of peptides with virtual tryptic digests of proteins encoded by ORFs in the genome of the organism (in this case *C. albicans*) from which the tagged protein has been purified. Discussion with the mass spectrometrist carrying out this part of the analysis is important, as many laboratories have their own "preferred" methods. As an alternative, it is also possible to digest the whole sample and carry out MALDI-TOF analysis of the material *en masse*. Then this requires rounds of filtering of the dataset to remove peptides corresponding to each protein as it is identified.

3.4.1. SDS-PAGE and Coomassie Staining

1. For initial analysis, a proportion of cell extracts, washes and elutions are examined. Prior to running samples on SDS-PAGE gels, it is usually necessary to concentrate the proteins in all but the initial clarified cell extract and unbound material after incubation with IgG-Sepharose^{TM}. This is conveniently achieved by trichloroacetic acid (TCA) precipitation.

2. An equal volume of 30% w/v TCA is mixed with each sample, and this is left on ice for 15 min before centrifuging at 15,000 × g for 15 min. The supernatant is removed, 0.5 mL acetone added to the tube, which is then vortexed thoroughly before repeating the centrifugation step, removing the acetone and allowing samples to air dry.

3. SDS-PAGE sample buffer is added and the samples denatured by incubation at 95°C and loading onto a 10% SDS-PAGE gel (this percentage is chosen as it resolves a broad range of proteins adequately). Typical volumes to analyse at this stage would be 5 µL of extract and unbound material from the IgG-Sepharose^{TM}, 250 µL of TEV eluate, 400 µL (i.e., an equal proportion) unbound material from the calmodulin affinity resin and 50–100 µL each elution from the calmodulin affinity resin. We do not attempt spectrophotometric quantification of the amount of material in each of the final elution fractions as this wastes valuable sample and cannot provide information on the success of the purification in terms of how intact proteins are, how complex the mixture is or any estimate to be made of the proportions of different proteins present in the sample. Bands that appear in sub-stoichiometric amounts may indicate that these factors interact transiently with the tagged protein, have weak interactions with it or that alternative complexes form with the tagged protein.

5. SDS-PAGE is carried out essentially as described *(25)*, using 20 × 20 cm, 1 mm thick gels (run in a Scie-Plas V20-CDC apparatus), electrophoresed at 150 V until the tracking dye (bromophenol blue) in the sample buffer is at the bottom of the gel. We use 20 cm long gels as they yield excellent resolution, but other formats and equipment can be used. Acrylamide gels 10% w/v are cast using 20 mL of resolving gel mix (comprising 5 mL resolving gel buffer, 5 mL 40% w/v acrylamide, 3 mL 2% w/v *bis*-acrylamide, 0.2 mL 10% w/v SDS, 6.5 mL water, 8 µL TEMED and 333 µL 10% w/v APS). This is overlaid with 1 mL water-saturated isobutanol until set (about 1 h). The isobutanol is poured off, the top of the gel rinsed with water, dried and the stacking gel poured on top (per 10 mL: stacking gel buffer, 1.25 mL 40% w/v acrylamide, 0.66 mL 2% w/v *bis*-acrylamide, 0.1 mL 10% w/v SDS, 6.4 mL water, 5 µL TEMED and 120 µL 10% w/v APS). The comb is then inserted and the gel again left to set for 1 h. Gels are then placed in the running apparatus

and 1X running buffer added to upper and lower chambers. Prior to loading SDS gel-loading buffer (50 mM Tris.HCl pH6.8, 100 mM dithiothreitol, 2% w/v SDS, 10% v/v glycerol and 0.1% w/v bromophenol blue tracking dye) is added to samples to at least 50% of the total volume and they are heated to 100°C for 5 min to denature proteins. It is important to vent the sample tubes or place something on top of them to prevent tubes from popping open during this step.

6. Following electrophoresis, the gel is stained using the Colloidal Blue Staining Kit according to manufacturer's instructions (*see* **Note 5**). At the scale of purification suggested here, we typically find that the tagged protein is readily visible with this method of staining (note that it will be larger than its predicted size due to the presence of the calmodulin-binding peptide retained through the purification). If the tagged protein is not seen or bands are faint, then the gel can subsequently be silver stained. Confirmation that the tagged protein has been purified can also be obtained by Western blotting a small proportion of the calmodulin affinity resin eluates. This requires that a suitable antibody is available to the protein of interest.

7. Stained bands can be cut directly from the stained gel for mass spectrometric analysis or, to increase the likelihood of success, a larger proportion of the eluates may be run on another gel. If the purified material is in several of the calmodulin affinity resin eluates, these can be pooled and concentrated, increasing the amount of material that can be run in a single gel lane.

3.5. Example

The power of the TAP-tagging approach to allow purification of complexes is aptly illustrated by the purification shown in **Fig. 13.2**. Signal recognition particle is a conserved ribonucleoprotein required for protein targeting to the endoplasmic reticulum. In most eukaryotes, this comprises six proteins (SRP9, SRP14, SRP19, SRP54, SRP68 and SRP72; named by the size of the mammalian components) plus an RNA. To characterise this complex in *C. albicans*, we used BLAST searches with *S. cerevisiae* sequences to identify genes putatively encoding several of its subunits. We then amplified a 306 base fragment of the putative *SRP72* ORF and inserted this into the *Xma*I and *Nsi*I sites of plasmid CIp-C-TAP (**Fig. 13.1A**), creating an in-frame fusion to the sequences encoding the TAP-tag. This plasmid was linearized using *Bst*EII, a unique site within the fragment of *SRP72* used and transformed into strain CAI8 ($\Delta ura3::imm434/\Delta ura3::imm434, ade2::hisG/ade2::hisG$) *(26)*. Following selection of a strain that expressed the TAP-tagged protein, SRP was purified from the strain using the protocol above, except that ribosomes were removed by centrifugation at $125,000 \times g$ (40,000 rpm in a Beckmann Type

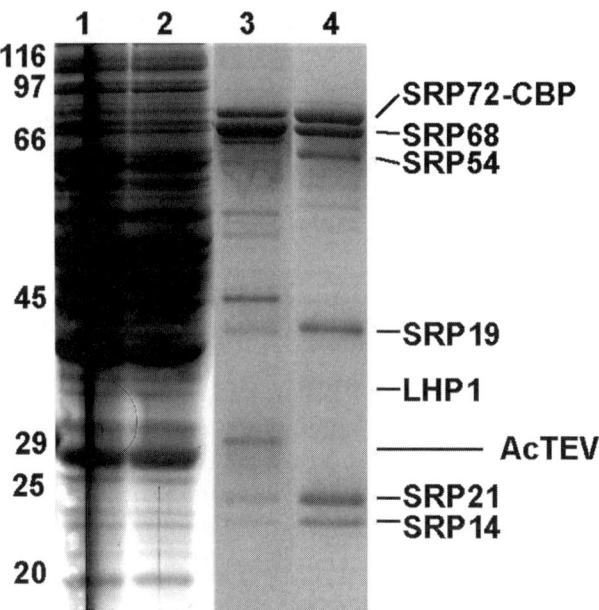

Fig. 13.2 Purification of *C. albicans* SRP. A *C. albicans* strain was generated which expressed TAP-tagged SRP72, by integration of a Clp-C-TAP derivative containing the 3' end of the SRP72 ORF (minus the stop codon) inserted into the polylinker. Post-ribosomal fractions of cell extract from the strain were incubated with IgG sepharose, which was washed and treated with AcTEV protease. The eluate was then incubated with calmodulin affinity resin and bound material eluted by removal of calcium ions with EGTA (see text for details). Proteins were resolved by SDS–PAGE and the gel stained with Colloidal Coomassie. Lane 1, extract; lane 2, unbound material; lane 3, TEV eluate; lane 4, calmodulin affinity resin eluate. Marker sizes are indicated to the left in kDa; note that while all lanes are from a single gel, a lighter image is shown for lanes 1 and 2 than lanes 3 and 4.

Ti45 rotor) for 170 min (sufficient to remove ribosomal subunits) and material was bound to IgG-Sepharose™ at 500 mM KOAc.

The single step from cell extract to IgG eluate revealed around 12 proteins when the fraction was resolved by SDS-PAGE and stained with Colloidal Coomassie (**Fig. 13.2**, Lane 3). Several of these were removed in the second step of the procedure on calmodulin-binding resin (Lane 4). One of the "contaminants" in the IgG column eluate at 29 kDa can be tentatively assigned as AcTEV™ protease, which can also be removed from purified fractions by immobilised metal ion chromatography due to the presence of a histidine-tag on this recombinant protein. Protein bands were excised from the gel, digested with trypsin and analysed by MALDI-TOF mass spectrometry. Spectra were then compared to a home-made database of translations of the predicted ORFs from assembly 19 of the *C. albicans* genome downloaded from the Stanford Genome Technology Center (http://www-sequence.stanford.edu/group/candida/). Apart from the smallest band, the principal proteins in the purified material were all shown to be SRP proteins. This left the complex without an SRP14 homologue, which was also absent from the available genome sequence. However, we were subsequently able to obtain primary amino acid sequence from the amino terminus of the purified protein, demonstrate that the smallest band was indeed the SRP14 homologue and that the ORF was missing from release 19 through errors in compiling the assembly (CB and JDB manuscript in preparation). Among the proteins that appeared to be present in less than stoichiometric quantities in the purification, the band at ~32 kDa was identified as the *C. albicans* La homologue. This protein binds to the 3' end of all nascent RNA polymerase III transcripts, of which the SRP RNA is one and has also been shown to co-purify with the *S. cerevisiae* SRP (e.g., *(27)*).

4. Notes

1. Blending can result in the production of aerosols of yeast, which can be a cause of concern in some circumstances. Alternative methods of lysing yeasts are in a French-press or a bead beater. We have successfully used a BeadBeater^{TM}, in conjunction with 0.5 mM zirconia/silica beads (both Biospec products, beads catalogue number 11079105z). Yeast "spaghetti" (up to 80 g) is thawed into lysis buffer as described (**Section 3.2**) in the ratio of 0.5 mL/g of cells, and this is transferred into the lysis chamber of the beater, together with ~200 mL of beads. The total volume of cells, lysis buffer and beads should not exceed ~350 mL as overfilling the lysis chamber with solid material causes the motor to overheat. Once the lysis chamber is mounted on the motor unit, ice mixed with salt (NaCl) is packed around the lysis chamber in the cooling jacket. Cells are lysed by the application of eight pulses, each lasting for 1 min, and separated by a cooling period of 2 min. The bead beater is disassembled and the liquid in the lysis chamber recovered. The beads remaining in the chamber are rinsed twice with 50 mL cold lysis buffer, which is pooled with the cell lysate. All the material collected is spun at $3000 \times g$ (e.g., 5000 rpm in a Beckmann JA20 rotor) for 10 min at 4°C. The supernatants from these tubes are pooled and centrifuged at $25,000 \times g$ (e.g., 15,000 rpm in JA20) for 15 min at 4°C. The cell debris and beads pelleted during first ($3000 \times g$) centrifugation are transferred back into lysis chamber still containing the majority of the beads, lysis buffer added to increase the volume to ~300 mL, and the lysis and harvesting steps repeated. The liquid phase is again collected in polycarbonate tubes but is immediately centrifuged at $25,000 \times g$ for

15 min at 4°C. The supernatants from both 25,000 × g spins are pooled and form the "total cell extract." It is important to keep all cell extracts at 4°C during the extraction process. Beads can be recycled by washing and sterilizing by baking in an oven at 160°C.

2. Various buffer conditions have been used in different laboratories for purification of TAP-tagged complexes. Alternative buffers (and protocols) to those suggested here, which have been used in systematic purification of multiple TAP-tagged proteins, can be found in *(14)* and at http://www-db.embl-heidelberg.de/jss/servlet/de.embl.bk.wwwTools.GroupLeftEMBL/ExternalInfo/seraphin/TAP.html. It is also important to consider factors that might affect the complex being purified, and which might affect the composition of the buffers used – i.e., it may be appropriate to develop application-specific buffers. Thus, for example, if it is suspected that protein–protein interactions within the complex being purified might be regulated by phosphorylation, phosphatase inhibitors (50 mM NaF, plus 2 mM Na_2Va_3) should be added to lysis and purification buffers.

3. Whether or not an extended incubation with IgG-sepharose resin is possible or, indeed, helpful, has to be empirically determined for each complex/protein being purified. Some fusion proteins are particularly susceptible to proteolysis and, similarly, some complexes are more stable than others. In the first instance, it is, therefore, advisable to limit the times of incubation with both IgG-sepharose and calmodulin affinity resins.

4. This step may also be performed at lower temperatures for longer periods (2–3 h at 16°C or overnight at 4°C), and the AcTEV™-buffer solution should be equilibrated to the appropriate temperature before use. As an alternative, the slurry of beads and buffer containing AcTEV™ can be transferred to a sealed tube at this stage and the incubation carried out on a roller. This ensures thorough mixing and for this reason can be more efficient than a static incubation.

5. There are a number of alternatives to Coomassie as used here. A commonly used alternative is Sypro-Ruby (Invitrogen), though this fluorescent dye (excitation and emission peaks at 450 and 610 nm, respectively) does require equipment suitable to visualise it. If silver staining is used, care must be taken in choosing an appropriate method, as several silver-staining protocols are incompatible with tryptic digestion and/or mass spectrometry due to chemical modification of proteins by aldehydes included in the procedures. Several compatible silver-staining protocols are discussed in *(28)*. When necessary we use the SilverQuest™ kit (Invitrogen), which is specifically formulated with mass spectrometry in mind.

Acknowledgements

The work described herein was funded by a UK Medical Research Council Senior Non-clinical Fellowship (to JDB). Thanks to Janet Quinn and Debbie Smith for comments on the manuscript.

References

1. Park, Y.N. and Morschhauser, J. (2005) Tetracycline-inducible gene expression and gene deletion in *Candida albicans*. *Eukaryotic Cell* **4**, 1328–1342.
2. Reuss, O., Vik, A., Kolter, R., and Morschhauser, J. (2004) The *SAT1* flipper, an optimized tool for gene disruption in *Candida albicans*. *Gene* **341**, 119–127.
3. Gerami-Nejad, M., Hausauer, D., McClellan, M., Berman, J., and Gale, C. (2004) Cassettes for the PCR-mediated construction of regulatable alleles in *Candida albicans*. *Yeast* **21**, 429–436.
4. Gerami-Nejad, M., Berman, J., and Gale, C.A. (2001) Cassettes for PCR-mediated construction of green, yellow, and cyan fluorescent protein fusions in *Candida albicans*. *Yeast* **18**, 859–864.
5. Care, R.S., Trevethick, J., Binley, K.M. and Sudbery, P.E. (1999) The *MET3* promoter: a new tool for *Candida albicans* molecular genetics. *Mol. Microbiol.* **34**, 792–798.
6. Kaneko, A., Umeyama, T., Hanaoka, N., Monk, B.C., Uehara, Y., and Niimi, M. (2004) Tandem affinity purification of the *Candida albicans* septin protein complex. *Yeast* **21**, 1025–1033.
7. Blackwell, C., Russell, C.L., Argimon, S., Brown, A.J., and Brown, J.D. (2003) Protein A-tagging for purification of native macromolecular complexes from *Candida albicans*. *Yeast* **20**, 1235–1241.
8. Doyle, T.C., Nawotka, K.A., Purchio, A.F., Akin, A.R., Francis, K.P., and Contag, P.R. (2006) Expression of firefly luciferase in *Candida albicans* and its use in the selection of stable transformants. *Microb. Pathog.* **40**, 69–81.
9. Corvey, C., Koetter, P., Beckhaus, T., Hack, J., Hofmann, S., Hampel, M., Stein, T., Karas, M., and Entian, K.D. (2005) Carbon Source-dependent assembly of the Snf1p kinase complex in *Candida albicans*. *J. Biol. Chem.* **280**, 25323–25330.
10. Backen, A.C., Broadbent, I.D., Fetherston, R.W., Rosamond, J.D., Schnell, N.F., and Stark, M.J. (2000) Evaluation of the *CaMAL2* promoter for regulated expression of genes in *Candida albicans*. *Yeast* **16**, 1121–1129.
11. Barelle, C.J., Manson, C.L., MacCallum, D.M., Odds, F.C., Gow, N.A., and Brown, A.J. (2004) GFP as a quantitative reporter of gene regulation in *Candida albicans*. *Yeast* **21**, 333–340.
12. Fromont-Racine, M., Rain, J.C., and Legrain, P. (1997) Toward a functional analysis of the yeast genome through exhaustive two-hybrid screens. *Nat. Genet.* **16**, 277–282.
13. Ito, T., Chiba, T., Ozawa, R., Yoshida, M., Hattori, M., and Sakaki, Y. (2001) A comprehensive two-hybrid analysis to explore the yeast protein interactome. *Proc. Natl. Acad. Sci. USA* **98**, 4569–4574.
14. Rigaut, G., Shevchenko, A., Rutz, B., Wilm, M., Mann, M., and Seraphin, B. (1999) A generic protein purification method for protein complex characterization and proteome exploration. *Nat. Biotechnol.* **17**, 1030–1032.
15. Ho, Y., Gruhler, A., Heilbut, A., Bader, G.D., Moore, L., Adams, S.L., Millar, A., Taylor, P., Bennett, K., Boutilier, K., et al. (2002) Systematic identification of protein complexes in *Saccharomyces cerevisiae* by mass spectrometry. *Nature* **415**, 180–183.
16. Gavin, A.C., Aloy, P., Grandi, P., Krause, R., Boesche, M., Marzioch, M., Rau, C., Jensen, L.J., Bastuck, S., Dumpelfeld, B., et al. (2006) Proteome survey reveals modularity of the yeast cell machinery. *Nature* **440**, 631–636.
17. Krogan, N.J., Cagney, G., Yu, H., Zhong, G., Guo, X., Ignatchenko, A., Li, J., Pu, S., Datta, N., Tikuisis, A.P., et al. (2006) Global landscape of protein complexes in the yeast *Saccharomyces cerevisiae*. *Nature* **440**, 637–643.

18. Forler, D., Kocher, T., Rode, M., Gentzel, M., Izaurralde, E., and Wilm, M. (2003) An efficient protein complex purification method for functional proteomics in higher eukaryotes. *Nat. Biotechnol.* **21**, 89–92.
19. Ihmels, J., Bergmann, S., Gerami-Nejad, M., Yanai, I., McClellan, M., Berman, J., and Barkai, N. (2005) Rewiring of the yeast transcriptional network through the evolution of motif usage. *Science* **309**, 938–940.
20. Enjalbert, B., Smith, D.A., Cornell, M.J., Alam, I., Nicholls, S., Brown, A.J., and Quinn, J. (2006) Role of the Hog1 stress-activated protein kinase in the global transcriptional response to stress in the fungal pathogen *Candida albicans*. *Mol. Biol. Cell* **17**, 1018–1032.
21. Tripathi, G., Wiltshire, C., Macaskill, S., Tournu, H., Budge, S., and Brown, A.J. (2002) Gcn4 co-ordinates morphogenetic and metabolic responses to amino acid starvation in *Candida albicans*. *EMBO J.* **21**, 5448–5456.
22. Delbruck, S. and Ernst, J.F. (1993) Morphogenesis-independent regulation of actin transcript levels in the pathogenic yeast *Candida albicans*. *Mol. Microbiol.* **10**, 859–866.
23. Brand, A., MacCallum, D.M., Brown, A.J., Gow, N.A., and Odds, F.C. (2004) Ectopic expression of URA3 can influence the virulence phenotypes and proteome of *Candida albicans* but can be overcome by targeted reintegration of *URA3* at the *RPS10* locus. *Eukaryotic Cell* **3**, 900–909.
24. Puig, O., Caspary, F., Rigaut, G., Rutz, B., Bouveret, E., Bragado-Nilsson, E., Wilm, M., and Seraphin, B. (2001) The tandem affinity purification (TAP) method: a general procedure of protein complex purification. *Methods* **24**, 218–229.
25. Laemmli, U.K. (1970) Cleavage of structural proteins during the assembly of the head of bacteriophage T4. *Nature* **227**, 680–685.
26. Fonzi, W.A. and Irwin, M.Y. (1993) Isogenic strain construction and gene mapping in Candida albicans. *Genetics* **134**, 717–728.
27. Gavin, A.C., Bosche, M., Krause, R., Grandi, P., Marzioch, M., Bauer, A., Schultz, J., Rick, J.M., Michon, A.M., Cruciat, C.M., et al. (2002) Functional organization of the yeast proteome by systematic analysis of protein complexes. *Nature* **415**, 141–147.
28. Mortz, E., Krogh, T.N., Vorum, H., and Gorg, A. (2001) Improved silver staining protocols for high sensitivity protein identification using matrix-assisted laser desorption/ionization-time of flight analysis. *Proteomics* **1**, 1359–1363.

Chapter 14

Preparation of Samples for Proteomic Analysis of the *Candida albicans* Cell Wall

Neeraj Chauhan

Abstract

Proteomics is the term for the large-scale analysis of proteins. Such studies have contributed greatly to our understanding of gene function in the postgenomic era. Besides identification, the characterization of protein function is also a key objective of proteomic research. One of the major challenges for *Candida* biologists is how to interpret the large amounts of transcriptional analysis data, and assign function(s) to numerous uncharacterized genes, and the proteins they encode. In this regard, the *Candida albicans* cell wall and its constituents have been a topic of research for many scientists largely due to its importance as a potential target for antifungal therapy and its role in pathogenesis. Thus, the application of proteomic technology to cell wall analysis is timely, and cell wall fractionation followed by identification of cell wall proteins by mass spectrometry is discussed in this chapter.

Key words: *Candida*, cell wall proteins, 2D-PAGE, mass spectrometry, MALDI-TOF/MS.

1. Introduction

The advent of proteomic technologies and the availability of completed genome sequence of *Candida albicans* has accelerated research concerning the biology of the organism. The cell wall of *C. albicans* has been shown to be an essential component to almost every aspect of its biology and pathogenicity. Furthermore, it is comprised of approximately 80–90% carbohydrate, which is a complex extracellular matrix of β-glucans, chitin, and mannan, whose composition changes during cell growth (*1–3*). Mannoproteins also constitute an important component of the cell wall, accounting for approximately 40% of total carbohydrate content. In order to better understand protein–function relationships of wall constituents, the application of methods for proteomic analysis is appropriate. Such analysis is two-part and the protocols described herein for mannoprotein analysis include (i) fractionation of the desired wall components (e.g., mannoproteins) are described, and (ii) analysis by two-dimensional PAGE in conjunction with mass spectroscopy (4–7).

Several different mass spectrometers are commercially available with varying degrees of sensitivity, such as, matrix-assisted laser desorption/ionization time-of-flight MS (MALDI-TOF MS). MALDI-TOF measures peptide mass only, while more sophisticated mass spectrometers determine peptide sequence as well. The main focus of this chapter is to outline methods available for proteomic analysis of the *Candida* cell wall and cell wall-associated proteins.

2. Materials

2.1. Organism and Medium

1. *C. albicans* wild-type strain SC5314 or CAF2-1 and any mutant strain for comparative study (*see* **Note 1**).
2. Yeast-extract–peptone–dextrose (YPD): 1% yeast extract, 2% peptone, and 2% dextrose.

2.2. Buffers and Stock Solutions

1. Lysis buffer: 10 mM Tris-HCl pH 7.5, 1 mM PMSF (*see* **Note 2**).
2. Extraction buffer: 50 mM Tris, pH 8.0, 2% SDS, 10 mM DTT, 0.1 M EDTA.

2.2.1. Cell Wall Protein Isolation and TCA Precipitation

3. Quantazyme (yeast lytic buffer) in 50 mM Tris-HCl, pH7.5 and 10 mM DTT.

2.2.2. Two-Dimensional PAGE

1. Lysis buffer: 7 M urea, 2 M thiourea, 2% CHAPS, 65 mM DTT, 0.5% carrier ampholytes, bromophenol blue.
2. First dimension: carry out first dimension using IPGphor system (Amersham Bioscience) following manufacturers instruction.
3. Second dimension, separating gel buffer (4X) stock: 1.5 M Tris-HCL, pH 8.8, 0.4% SDS. Filter the solution through a 0.45-µm filter and store at room temperature for up to a month. Separating gel stock solution: 40% acrylamide/2.5% piperazine diacrylamide (*see* **Note 3**). Filter the solution through a 0.45-µm filter and store at 4°C in the dark. Discard after a month.
4. Ammonium persulfate (Bio-Rad): 25% (w/v) solution in distilled water. Prepare fresh.

5. TEMED (Bio-Rad).
6. Gel running buffer (5X): 15.1 g Tris base (0.125 M), 72 g glycine (0.96 M), 5 g SDS (0.5%), DW to 1 L; pH of the stock is 8.3 when diluted. Store at 4°C for up to 1 month.
7. Transfer buffer (1X): 3.03 g Tris base (24 mM), 14.4 g glycine (194 mM), 200 mL methanol (20%), DW to 1 L.
8. Sample loading buffer (6X) stock: 4X Tris-HCl/SDS pH 6.8, SDS (10% final conc.), glycerol (30% final), DTT (0.6 M final), bromophenol blue (0.012% final), add DW to 10 mL (if needed).
9. Tris buffered saline/Tween (TBS-T): 25 mM Tris-HCL pH 8.0, 125 mM NaCl, 0.1% Tween 20.

3. Methods

3.1. Strains and Growth Conditions

1. *C. albicans* strains are grown overnight in YPD at 30°C. Overnight cultures are then diluted in fresh YPD medium to OD_{600} ~0.1 and grown until cells reach log phase.

3.2. Isolation of Cell Wall Protein

1. Grow cells overnight at 30°C.
2. Inoculate 2 L YPD liquid medium in a large flask and grow until OD_{600} ~1.0 (log phase).
3. Harvest cells by centrifugation at 3500 × *g* for 5 min.
4. Wash cell pellet with cold distilled water. At this stage, cells can be stored at −80°C until further use.
5. Wash cells five times with lysis buffer. After washing, resuspend cells in ice-cold lysis buffer (*see* **Note 4**).
6. Add equal volume of glass beads (0.45 μm, Sigma) and transfer cells to a bead beater (biospec products). Bead beat cells with 30 s pulse for 10 min at 4°C (*see* **Note 5**).
7. Centrifuge at 3000 × *g* for 10 min.
8. If not used immediately, store supernatant (cytoplasmic fraction) at −80°C until analysis.
9. Wash cell wall fraction (pellet) 5X with ice-cold distilled water. Wash 5X with each of the following ice-cold solutions in sequence: 5% NaCl, 2% NaCl, 1% NaCl containing 1 mM PMSF, and finally, wash 1X with ice-cold distilled water.
10. Extract cell wall proteins twice by boiling in extraction buffer for 10 min.

11. Following two extractions, wash wall 5X with ice-cold water and then a further 10X with 0.1 M sodium acetate, pH 5.5 containing 1 mM PMSF.

12. Divide cell pellet into two. Extract fraction 1 with NaOH (30 mM) overnight at 4°C. Digest fraction 2 with quantazyme (yeast lytic enzyme) at 37°C overnight.

13. Precipitate cell wall proteins with TCA.

3.3. TCA Precipitation

1. Add sodium deoxycholate (20 mg/mL stock, >pH 8) to give a sample concentration of 80–200 µg/mL.
2. Incubate on ice for 30 min.
3. Add TCA (20% stock) for a final concentration of 6%.
4. Incubate on ice for 90 min.
5. Centrifuge at 10,000 × g for 20 min at 4°C.
6. Carefully remove supernatant.
7. Wash sample with acetone (chilled to –20°C).
8. Centrifuge at 10,000 × g for 20 min at 4°C.
9. Carefully remove acetone.
10. Quantitate protein by using Bradford protein assay method (BioRad).
11. Store samples at –80°C.

3.4. Two-Dimesional PAGE Analysis

1. Solubilize protein samples, 200–500 µg for analytical gels and 5–10 mg for preparative gels in lysis buffer and applied to 18 cm immobiline dry strips pH 3–10 (Amersham BioScience).
2. Perform IEF following manufacturer's protocol (*see* **Note 6**).
3. Equilibrate sample strips in 50 mM Tris-HCl, pH 8.8, containing 6 M urea, 30% v/v glycerol, 2% SDS, 135 mM iodoacetamide (*see* **Note 7**).
4. For second dimension, a stacking gel is not needed because the proteins are preseparated by IEF and migrate from one gel to another. Carry out electrophoresis on 1.5 mm thick 10%T, 1.6% C (piperazine diacrylamide) PAGE gel (*see* **Note 8**) at 5 mA for 2 h and then for 15 mA overnight. By the end of electrophoresis, bromophenol blue tracking dye should migrate to the end of the gel.

3.5. Western Blot

Western blots are performed after cell wall fractionation and two-dimensional PAGE or SDS–PAGE analysis using specific antibodies against cell wall proteins, following standard protocols (*8*) and as detailed below.

1. Proteins from analytical two-dimensional gels or SDS–PAGE are transferred to nitrocellulose membrane (BioRad) using transfer buffer at 50 mA overnight in cold room.
2. Incubate nitrocellulose membrane in 5% nonfat dry milk for 1 h at room temperature.
3. Wash membrane three times for 5 min with TBS-T.
4. Incubate with primary antibody (at the appropriate dilution) for 1 h at room temperature (*see* **Note 9**).
5. Wash membrane three times for 5 min with TBS-T.
6. Incubate membrane with secondary antibody for 1 h at room temperature and develop with ECL detection kit (Amersham Biosciences) following manufacturer's instructions.

3.6. Gel Staining and Image Analysis

1. Two-dimensional analytical gels are silver stained (*see* **Note 10**) or preparative gels are stained with Coomassie Brilliant Blue for visualizing proteins. Gel images are then scanned and quantitated using a densitometer.
2. Protein spots of interest are excised from gel for identification by mass spectrometry.

3.7. Mass Spectrometric Analysis

The proteins separated by gel electrophoresis are identified by matrix-assisted laser desorption/ionization time-of-flight MS (MALDI-TOF MS). Prior to MALDI-TOF, MS protein spots are excised from gel and destained as follows:

3.7.1. Sample Preparation

1. Incubate excised protein in 100 mM sodium thiosulfate and 30 mM potassium ferricyanide.
2. Wash with 25 mM ammonium bicarbonate followed by washing with water.
3. Incubate in 100% acetonitrile for 15 min.
4. Dry in SpeedVac for 15 min.

3.7.2. Reduction, Alkylation, and Digestion of Protein Samples

1. Destained proteins are reduced with 10 mM DTT in 25 mM ammonium bicarbonate at 56°C for 30 min.
2. Reduced protein is alkylated with 55 mM iodoacetamide in dark for 15–20 min with 25 mM ammonium bicarbonate.
3. For digestion, incubate all protein samples with 12.5 ng/μL sequencing grade trypsin (Roche) overnight at 37°C with 25 mM ammonium bicarbonate.
4. Separate crude extracts (supernatant) by centrifugation at $5000 \times g$ for 5 min.
5. Extract peptides with 50% acetonitrile, 1% trifluoroacetic acid, and finally with 100% acetonitrile. Pool all the extracts together and concentrate by SpeedVac.

3.8. MALDI-TOF-MS Concentrated protein samples obtained after digestion are now ready for downstream analysis using MALDI-TOF MS.

3.9. Database Search Perform protein sequence search with data obtained after two-dimensional and mass spectroscopy analysis. Searches can be performed at www.candidagenome.org and www.expasy.ch/sprot, among other sites.

4. Notes

1. Start overnight culture from a fresh colony, and maintain *Candida* cultures on fresh YPD plates.
2. Use only MilliQ grade water for preparation of all buffers and stock solutions.
3. Unpolymerized acrylamide is neurotoxic. Utmost care should be taken to avoid exposure. Wear protective gear (gloves, mask, etc.) while handling.
4. Keep all solutions on ice during cell wall protein extraction. It is important to add 1 mM PMSF to avoid protein degradation.
5. Cell lysis by glass beads should be verified by microscopy.
6. It is recommended to equilibrate protein samples immediately after the first dimension or alternatively, IPG strips can be stored at -80^0C until needed.
7. Prepare fresh urea solution to avoid formation of isocyanate (it may lead to vertical streaks).
8. Ensure that the contact between first and second dimensional gels is free of air bubbles. Presence of bubbles will cause horizontal streaks. The total monomer concentration in the gel is denoted as %T and the percentage of total monomer (cross-linker), e.g., piperazine diacrylamide, is denoted as %C. Second-dimension gels are of two types: homogeneous gels, with constant %T and %C, and gradient gels, with increasing %T, usually with constant %C. The choice of %T is determined by the molecular weight of the protein to be separated. Typically, a homogeneous 10–12% or gradient 8–18% gel is best for crude samples, such as a whole cell lysate.
9. Primary antibody can be reused for subsequent experiments by adding 0.02% sodium azide and stored at 4°C for up to a month.
10. For optimal silver staining, it is essential to use high-quality reagents and to prepare the solutions fresh each time. It is also recommended to use glass trays during staining because

this results in less background staining than plastic trays. For gels cross-linked with piperazine diacrylamide, use silver diamine stain.

References

1. Chauhan, N., Li, D., Singh, P., Calderone, R., and Kruppa, M. (2002) The cell wall of *Candida* spp. In: *Candida and Candidiasis* (Richard A. Calderone, ed.), ASM Press, Washington DC, pp. 159–175.
2. Chaffin, W. L., Lopez-Ribot, J. L., Casanova, M., Gozalbo, D., and Martinez, J. P. (1998) Cell wall and secreted proteins of *Candida albicans*: identification, function, and expression. *Microbiol. Mol. Biol. Rev.* **62**, 130–180.
3. Kapteyn, J. C., Hoyer, L. L., Hecht, J. E., Muller, W. H., Andel, A., Verkleij, A. J., Makarow, M., Van Den Ende, H., and Klis, F. M. (2000) The cell wall architecture of *Candida albicans* wild-type cells and cell wall-defective mutants. *Mol. Microbiol.* **35**, 601–611.
4. de Groot, P. W., de Boer, A. D., Cunningham, J., Dekker, H. L., de Jong, L., Hellingwerf, K. J., de Koster, C., and Klis, F. M. (2004) Proteomic analysis of *Candida albicans* cell walls reveals covalently bound carbohydrate-active enzymes and adhesins. *Eukaryotic Cell* **3**, 955–965.
5. Ebanks, R. O., Chisholm, K., McKinnon, S., Whiteway, M., and Pinto, D. M. (2006) Proteomic analysis of *Candida albicans* yeast and hyphal cell wall and associated proteins. *Proteomics* **6**, 2147–2156.
6. Pitarch, A., Sanchez, M., Nombela, C., and Gil, C. (2002) Sequential fractionation and two-dimensional gel analysis unravels the complexity of the dimorphic fungus *Candida albicans* cell wall proteome. *Mol. Cell. Proteomics* **1**, 967–982.
7. Pitarch, A., Jimenez, A., Nombela, C., and Gil, C. (2006) Decoding serological response to *Candida* cell wall immunome into novel diagnostic, prognostic, and therapeutic candidates for systemic candidiasis by proteomic and bioinformatic analyses. *Mol. Cell. Proteomics* **5**, 79–96.
8. Chauhan, N., Inglis, D., Roman, E., Pla, J., Li, D., Calera, J. A., and Calderone, R. (2003) *Candida albicans* response regulator gene *SSK1* regulates a subset of genes whose functions are associated with cell wall biosynthesis and adaptation to oxidative stress. *Eukaryotic Cell* **2**, 1018–1024.

Chapter 15

Reporter Gene Assays in *Candida albicans*

Joy Sturtevant

Abstract

Reporter systems are used in *Candida albicans* in three major experimental areas. These include gene expression, promoter analysis, and protein expression/localization. Heterologous expression in *C. albicans* is either not effective or inefficient due to the alternative codon usage in *Candida*, particularly CTG. Consequently, several reporter genes have been constructed by optimizing codons for expression in *Candida*. The reporter systems include lacZ, luciferase, and GFP. Generally, PCR site directed mutagenesis has been used to construct the modified reporter. Reporter gene vectors are not commercially available for *Candida*, but they can normally be requested from the laboratories that developed the constructs.

Key words: *Candida*, reporter assays, GFP, luciferase, lacZ, gene expression, promoter analysis.

1. Introduction

Three reporter systems have been adapted for use with *Candida albicans*. These include lacZ *(1)*, luciferase *(2, 3)*, and GFP *(4–7)*. All three have been designed to study gene expression and are amenable for promoter analysis. In addition, GFP constructs have been designed to study protein localization *(5, 6)*. The reporter genes are either ligated 3' of a target promoter or fused in frame to the NH_2 or COOH terminal of the target protein.

β-Galactosidase is a popular reporter system in model systems. Transformants are easily visualized using a white-blue selection, and quantification of expression can be assessed by a liquid assay. Although *C. albicans* has no endogenous β-galactosidase gene, nor can it utilize lactose as a carbon source, a reliable, sensitive system has recently been reported. The first β-galactosidase system for *C. albicans* was constructed by modifying the *lacZ* gene from *Kluyveromyces lactis* and ligating it into a multicopy plasmid *(8)*.

However, this construct had the inherent disadvantage of plasmid instability *(9, 10)*. Uhl and Johnson ligated the codon-modified *lacZ* genes from both *K. lactis* and *Streptococcus thermophilus* into an integrative vector and placed the lacZ under the inducible *MAL2* promoter *(1)*. They were able to detect significant activity with the *S. thermophilus* but not the *K. lactis* lacZ. Additionally, the *S. thermophilus* lacZ CTG codon did not have to be modified and was successfully used with additional promoters (*ACT1* and *HWP1*) *(1)*. The lacZ reporter system has been used to study the regulation of the histidine kinase gene, *CHK1*, and promoter analysis of *CDR1* *(11, 12)*. A lacZ system has also been designed for *C. glabrata* *(13)*.

Two reporter luciferase systems have been designed for *C. albicans*. The first system was developed using the luciferase gene (*RLUC*) from the sea pansy *Renilla reniformis* since it contains no CTG codons *(3)*. *RLUC* was ligated 3' of several promoters and integrated into the *ADE1* locus of an *ade1* derivative of strain WO1. The sea pansy system was used to perform promoter analysis in white opaque switching transition in strain WO1 *(14)*. This system can be modified for the target strain and promoter of choice. A second luciferase reporter system was designed using firefly luciferase *(2)*. The luciferase expression construct (pGTV-ENO) has been CTG codon modified and contains several additional features: *(1)* the 9 CTG codons were modified. *(2)* A *C. albicans* ADH termination sequence was ligated directly 3' The constitutive enolase promoter was ligated directly 5' of the luciferase gene. Restriction sites allow easy exchange of promoter sequences. Gene-targeting sequences flanking the luciferase expression cassette are designed so that the luciferase gene integrates into an intergenic region in chromosome 1 that does not interfere with any tested *C. albicans* phenotypes. The plasmid also contains the *URA3* gene though it is not required for selection of transformants. Consequently, clinical strains can be transformed *(2)*.

The GFP reporter system is the most widely used in *C. albicans*. GFP systems in *Candida* are used to assess gene expression, promoter analysis, and protein localization *in vitro* and *in vivo*. Cormack et al. systematically modified every codon in the *Aequorea victoria* GFP and incorporated two mutations that enhance fluorescence to construct a yeast-enhanced GFP (yEGFP3) *(4)*. yEGFP3 was ligated into the autonomously replicating plasmid YPB-ADHpt . Morschhauser was able to make a GFP expression construct by only altering the CTG codons *(7)*. This construct is stably integrated into the *ACT1* locus, and the selectable marker is *URA3*. Both of these constructs require conventional cloning techniques for expression in *C. albicans*. Additional GFP integrative systems have been constructed by other laboratories. Staab *et al.* constructed hyphal-GFP expression integrative vectors (pHWP1GFP3) *(15)*. The plasmid contains the *HWP1* promoter upstream of yEGFP3 (from Cormack's original vector), *URA3* to complement strains CAI4 or BWP17, and *ENO1* sequence so the construct integrates into the *ENO1* locus *(15)*. Barelle *et al.* modified the original vector designed by Cormack by introducing yEGFP into an *RPS1* integrative vector *(16)*. This construct has been used successfully *in vivo (17)*. A larger set of constructs is now available that allows one-step PCR construction of fluorescent protein fusion constructs *(5, 6)*. These vectors are designed so the FP tag is fused 3' of the target gene by homologous recombination. An additional set is designed for fusion 5' of the target gene, which is then expressed under the control of regulatable promoters *(6)*. Vectors are also available with yellow fluorescent protein (YFP) and cyan fluorescent protein (CFP) *(5)*. Mao *et al.* used GFP to study the amino and carboxy termini peptide signaling of GPI-anchored proteins' transport to the cell wall. These vectors allow fusion to either amino or carboxy termini and utilize the *TEF2* promoter *(18)*. GFP reporter systems have also been used successfully in other *Candida* species including *C. boidinii (19)*, *C. glabrata (20, 21)*, and *C. dubliensis (22)*.

2. Materials

2.1. Construction of Gene-Specific Reporter Constructs

Required materials are described chapters 16, 17. Recipient *C. albicans* strains with appropriate auxotrophic strains are required for constructs containing *URA3*, *ADE1*, and *HIS1*.

1. Clinical isolates can be used when constructs contain a SAT marker.
2. Primers must be designed with gene-specific sequence and/or appropriate restriction enzyme sites.
3. Agar plates with appropriate selective media.

2.2. β Glactosidase Assays

1. X-Gal: 20 mg/mL in dimethylformamide (DMF).
2. Plates: YNB + X-Gal (2 mL of 20 mg/mL X-Gal/L.)
3. X-Gal modified medium (XMM) – 1L:1.7 g yeast nitrogen base w/o amino acids or ammonium sulfate, 20 g glucose, 5 g ammonium sulfate, 20 g agar, Add to 930 mL H20.
4. Autoclave.
5. Add 70 mL 1 M potassium phosphate, pH 7.0, and 2 mL of 20 mg/mL X-Gal solution.

6. Z buffer (500 mL): 4.26 g Na_2HPO_4 (anhydrous) (60 mM), 2.40 g NaH_2PO_4 (60 mM) 373 mg KCl (10 mM), 123 mg $MgSO4$ (1 mM), Adjust to pH 7.0.
7. *o*-Nitrophenyl-α-D–galactopyranoside (ONPG) (Sigma), 4 mg/mL.
8. Chloroform.
9. 0.1% SDS.
10. 1 M Na_2CO_3.
11. Liquid nitrogen.

2.3. Luciferase Assays

2.3.1. *Renilla reniformis* System

1. RLUC buffer: 0.5 M NaCl, 0.1 K_2HPO4, pH 6.7, 1 mM Na_2EDTA, 0.6 mM sodium azide.
   (a) 1 mM phenylmethylsulfonyl fluoride (PMSF), 0.02% bovine serum albumin.
   (b) Flush an aliquot of RLUC buffer with nitrogen (to prevent auto-oxidation of coelentraxine).
2. Coelentrazine (molecular probes).
   (a) Prepare stock solution in acid–methanol and store at –20°C.
   (b) Working solutions: dilute stock solution in nitrogen-flushed RLUC buffer so that final concentrations are 1 µM (microtiter assay) and 10 µM.
3. Coel-RLUC buffer: 0.5 µM in nitrogen-flushed RLUC buffer.

2.3.2. *Photinus pyralis* Luciferase (Firefly) System

1. Luciferin (Biosynth). Prepare 10 and 50 µg/mL solutions.
2. Minimal (SD) media.

3. Methods

3.1. lacZ Reporter System (1)

3.1.1. Construction of lacZ Expressing Strain

1. Construct a *LACZ* transformation vector using the *S. thermophilus lacZ* gene (*see* **Note 1**). Possible vectors are constructed with standard molecular cloning methods. Vectors should be chosen as appropriate to the experiment with consideration to cloning sites and selection parameters.
2. Ligate the *LACZ* gene 3' of the promoter of choice (*see* **Note 2**).
3. Transform the *C. albicans* strain of choice with the *LACZ* construct and select transformants as described in other chapters in this volume (Chapters 16 and 17) (*see* **Note 3**).

3.1.2. Visual Identification of LacZp Expressing Strains

1. After transformation, patch transformants on YNB + X-Gal (*see* **Note 4**).

2. Include/exclude appropriate (i) amino acids dependent on the *C. albicans* strain; (ii) nutrient sources when LacZp is expressed under inducible/repressible promoters; and/or (iii) supplements (e.g., 10% fetal calf serum) if testing for expression during hyphal induction.

3. Positive colonies will turn from white to blue. The time depends on the promoter. Strains transformed with *ACT1-lacZ* turn blue within 2 days, while under the *MAL2-lacZ* takes 4–5 days on inducible media (*see* **Notes 5** and **6**).

4. In order to assess blue color more quickly, filter assays can be performed.
   (a) Prepare plates with colonies to be tested. Colonies should be 2–4-days old and 1–3 mm in diameter.
   (b) For each plate to be assayed, prepare a Whatman #5 or VWR 410 75 mm filter. Place filter in 3 mL of Z buffer with 20 µL of 2% X-Gal (freshly prepared) in a clean 100-mm plate.
   (c) Place a clean, dry filter over the plate of patched colonies of *C. albicans*.
   (d) Orient filter to agar plate by poking holes through filter in asymmetric pattern.
   (e) Once the filter is evenly wetted (colonies transferred to filter), carefully lift the filter and transfer (colony side up) to a beaker of liquid nitrogen.
   (f) Submerge filters in liquid nitrogen to crack colonies (lyse cell walls) for 10 s, repeat.
   (g) Place filter colony side up on preincubated filter.
   (h) Incubate 30 min to 8 h until color appears.
   (i) Identify β-galactosidase-producing colony by aligning the filter to the agar plate.

3.1.3. Quantification of LacZp Expression (24)

1. Resuspend 1 mL of cells in 1 mL of Z buffer and place on ice.
2. Determine OD600.
3. Resuspend 10–100 µL of cells in 1 mL Z buffer so OD600 ~1.0.
4. Permeabilize cells with 15 µL of 0.1% SDS and 30 µL chloroform.
5. Incubate cells at 37°C for 5 min.
6. Add 0.2 mL ONPG and incubate at 37°C until desired color (pale yellow) is obtained.
7. Reactions are stopped by addition of 0.5 mL of 1 M Na_2CO_3.

8. Centrifuge reaction at 10,000 x g for 5 min.
9. Determine A420 and A550 of supernatant.
10. Calculate units of activity:
    β-galactosidase units = $1000 \times OD420/t \times V \times OD600$,
    where t = time of incubation
    V = dilution of cells
    OD600 = A600 of 1 mL of cells (**Step 3**)

3.2. Luciferase Reporter Systems

Several reporter systems utilizing luciferase are available, and some of these methods are described below.

1. Prepare luciferase-expressing strains as above (**Section 3.1.1**) except use *RLUC* as the source of your reporter gene. Confirm appropriate integration by Southern blot analysis.

3.2.1. Renilla Reniformis Luciferase (Sea Pansy) After Srikantha et al. (3)

2. *In vitro* assay of RLUC activity
   (a) Grow cells under desired test conditions.
   (b) Wash cells (2×10^6) twice with sterile water and one time with RLUC buffer.
   (c) Resuspend cells in 200 µL of RLUC buffer; add two volumes of glass beads and disrupt with bead beater (four cycles of 20 s duration at 4°C). Number of cycles and duration will depend on bead beater.
   (d) Centrifuge extracts at $13,000 \times g \times 15$ min at 4°C.
   (e) Prepare serial dilutions of cell extracts. Usually between 1/1,000 and 1/10,000 are in the measurable range.
   (f) Determine protein concentration.
   (g) Add 1–2 µL of cell extract dilutions to 100 µL Coel-RLUC buffer in 4 mL tubes for light measurement, mix.
   (h) Immediately read in a luminometer at 480 nm in integration mode for 10–30 s.
   (i) Calculate activity as relative luminescence/10 s/µg of protein.

3. *In vivo* assay of RLUC activity
   (a) Grow cells under desired test conditions.
   (b) Wash twice with distilled water and once with RLUC buffer.
   (c) Resuspend cells at 2×10^8/mL.
   (d) Mix 50 µL of cells (10^7) with 50 µL of 10 µM coelentrazine.
   (e) Record luminescence for 30 s in integrative mode.
   (f) Calculate activity as relative luminescence/10 s/10^7 cells.

4. Microtiter plate assays
   (a) Serially dilute (10-fold) cells from **Step 3c** above (2×10^8 to 2×10^2/mL).
   (b) Add 50 μL of each suspension to a microtiter plate.
   (c) Add 50 μL of 1 μM coelentrazine, mix.
   (d) Place a piece of X-ray film firmly at the bottom of the plate and cover with aluminum foil.
   (e) Expose film for 6 h at 25°C.

3.2.2. *Photinus pyralis* Luciferase (Firefly) Reporter System After Doyle et al. *(2)*

1. Transform *C. albicans* with pGTV-ENO (or modified with promoter of choice).
2. Dilute transformation reaction to ~2.5×10^6 cells/mL. Plate 100 μL transformants on minimal media with 50 μg/mL luciferin and incubate at 30°C (see **Notes 7** and **8**).
3. Image plates using an imager with a cooled CCD camera and identify regions of bioluminescence. At this point, it is not possible to identify single colonies.
4. Collect patches with bioluminescent colonies and inoculate 1 mL SD cultures. Grow overnight shaking at 30°C.
5. Transfer 75 μL of overnight cultures to a black 96-well microtiter plate and add 1 μL of luciferin (16 mg/mL).
6. Image microtiter plate on imager and identify wells with highest light output.
7. Dilute these wells so that approximately 100 cells are plated on SD + luciferin plates. Incubate overnight at 30°C.
8. Image plates on imager and select isolated bioluminescent colonies.

3.3. GFP Reporter Systems

3.3.1. GFP Gene Expression Reporter Systems

1. Construct GFP integrative cassette with target promoter (*see* **Note 9**).
2. Construct GFP expression strains using routine procedures (*see* **Note 10**).
3. Prepare cells for fluorescent microscopy and observe under fluorescent microscope using appropriate filter sets.
4. Alternative: prepare cell extracts and subject to immunoblotting with commercially available anti-GFP antibodies.

3.3.2. GFP Protein Expression Reporter System from Gerami-Nejad et al. *[5]*

1. Select appropriate vector. Vectors are available with GFP, YFP, and CFP and the *HIS1* or *URA3* auxotrophic marker (*see* **Note 11**).
2. Design primers (*see* **Note 12**):
   (a) Forward primer. 70 nt just upstream of the stop codon plus the tag sequence: 5'-GGT GGT GGT TCT AAA

GGT GAA GAA TTA TT-3'. The tag sequence includes a glycine linker and is homologous to the 5' end of the PCR template. Gene-specific sequence must be in frame with tag sequence GGT/GGT.

Reverse primer. 70 nt gene-specific sequence just downstream of the stop codon plus 5'-TCTAGAAGGACCACCTTT-GATTG-3' for *URA3* or 5'-GAATTCCGGAATATTTAT-GAGAAAC-3' for *HIS1*. The latter sequences are complementary to the 3' end of the selectable marker genes in the PCR template.

3. Perform PCR and transform *C. albicans* with PCR products using PCR-mediated gene disruption (Chapter 17, this volume).
4. Confirm integration by PCR and Southern hybridization (*see* **Note 13**).
5. Prepare cells for fluorescent microscopy and observe under fluorescent microscope using appropriate filter sets.
6. Alternative: prepare cell extracts and subject to immunoblotting with commercially available anti-GFP antibodies.

3.3.3. Conditional GFP Fusion Proteins from Gerami-Nejad *et al.* (6)

1. Select appropriate vector: vectors are available with the *MET3*, *GAL1*, and *PCK1* regulatable promoters.
2. Design primers.
    (a) Forward primer: 70 nt gene-specific sequence immediately upstream of the gene of interest plus 5'–TCTAGAAG-GACCACCTTTGATTG-3', which is homologous with 5' end of *URA3* gene in PCR template.
    (b) Reverse primer: 70 nt that are complementary to the 5'; end of the gene of interest plus 5'; – TTT GTA CAA TTC ATC CAT AC -3'; The gene-specific interest must be in frame with TTT/GTA.
3. Continue as described in **Steps 3 and 4** above (**Section 3.3.2**).
4. Expression of GFP fusions are induced by growth in inducing media dependent on the chosen regulatable promoter.
5. GFP fusion proteins can be detected as described above (**Steps 5 and 6**, 3.3.2) (*see* **Note 14**).

4. Notes

1. It is necessary to integrate the construct containing the reporter gene into the chromosome due to the instability of plasmids in *C. albicans* (9, 10). However, it is important not to disrupt normal cellular processes including growth, differentiation, virulence, and cell metabolism. Several loci have been

identified, which fit these criteria and are used routinely. These include *RPS1* and *ACT1 (7, 25)*. Alternatively, constructs can be integrated in a manner that allows regeneration of native loci such as *URA3 (23)*.

2. The modified *lacZ* gene is present in plasmid pAU36 and can be obtained from Dr. A. Johnson *(1)*.

3. Expression of LacZp does not allow *C. albicans* to grow in the presence of lactose as the only carbon source *(1)*.

4. If necessary enhancement of β-galactosidase activity can be obtained by plating on XMM plates, but *C. albicans* grows more slowly on this media *(1)*.

5. Some transformants exhibit higher β-galactosidase activity. This is most likely due to multiple integrations *(1)*. These strains should not be used for further studies. For further comparative studies, correct integrations should be confirmed by Southern blotting.

6. Plate assays also allow identification of gene expression changes within a colony, e.g., genes expressed upon starvation, filamentation, etc.

7. The growth of transformed clinical isolates is not compromised, so colonies will grow as a lawn after 1 day. If transformation requires complementation of auxotrophy, then do not dilute cells and plate on a selective medium.

8. The firefly luciferase expression construct allows the integration of a reporter system in clinical strains. Consequently multiple selection steps have to be performed to isolate single colonies. This could be modified by including a drug selectable marker.

9. A variety of template plasmids are available that allow integration into different loci including *ACT1 (7)*, *ENO1 (15)*, and *RPS1 (16)*.

10. Several laboratories have successfully used the GFP reporter system for gene expression and promoter analysis *(4,7,15,17,26–28)*.

11. These constructs have the advantage in that they allow protein localization and expression under the control of endogenous promoter. These are only useful if the carboxy termini are not required for protein function.

12. The same forward primer can be used with all three fluorophores; a separate reverse primer is required for *URA3* and *HIS1*.

13. One primer is designed to anneal to the transformation cassette; the second primer is designed to anneal a region in the target gene that is not included in the PCR construct.

14. Care must be taken in interpreting the results from the conditional GFP fusion proteins since results will not always be physiologically applicable.
15. The GFP reporter system has an advantage over the lacZ and luciferase systems in that no substrates or additional reagents are needed.

References

1. Uhl, M. A., and Johnson, A. D. (2001) Development of *Streptococcus thermophilus lacZ* as a reporter gene for *Candida albicans*. *Microbiology* **147**, 1189–1195.
2. Doyle, T. C., Nawotka, K. A., Purchio, A. F., Akin, A. R., Francis, K. P., and Contag, P. R. (2006) Expression of firefly luciferase in *Candida albicans* and its use in the selection of stable transformants. *Microb. Pathog.* **40**, 69–81.
3. Srikantha, T., Klapach, A., Lorenz, W. W., Tsai, L. K., Laughlin, L. A., Gorman, J. A., and Soll, D. R. (1996) The sea pansy *Renilla reniformis* luciferase serves as a sensitive bioluminescent reporter for differential gene expression in *Candida albicans*. *J. Bacteriol.* **178**, 121–129.
4. Cormack, B. P., Bertram, G., Egerton, M., Gow, N. A., Falkow, S., and Brown, A. J. (1997) Yeast-enhanced green fluorescent protein (yEGFP)a reporter of gene expression in *Candida albicans*. *Microbiology* **143**, 303–311.
5. Gerami-Nejad, M., Berman, J., and Gale, C. A. (2001) Cassettes for PCR-mediated construction of green, yellow, and cyan fluorescent protein fusions in *Candida albicans*. *Yeast* **18**, 859–864.
6. Gerami-Nejad, M., Hausauer, D., McClellan, M., Berman, J., and Gale, C. (2004) Cassettes for the PCR-mediated construction of regulatable alleles in *Candida albicans*. *Yeast* **21**, 429–436.
7. Morschhauser, J., Michel, S., and Hacker, J. (1998) Expression of a chromosomally integrated, single-copy GFP gene in *Candida albicans*, and its use as a reporter of gene regulation. *Mol. Gen. Genet.* **257**, 412–420.
8. Leuker, C. E., Hahn, A. M., and Ernst, J. F. (1992), Beta-galactosidase of *Kluyveromyces lactis* (Lac4p) as reporter of gene expression in *Candida albicans* and *C. tropicalis*. *Mol. Gen. Genet.* **235**, 235–241.
9. Kurtz, M. B., Cortelyou, M. W., Miller, S. M., Lai, M., and Kirsch, D. R. (1987) Development of autonomously replicating plasmids for *Candida albicans*. *Mol. Cell. Biol.* **7**, 209–217.
10. Pla, J., Perez-Diaz, R. M., Navarro-Garcia, F., Sanchez, M., and Nombela, C. (1995). Cloning of the *Candida albicans HIS1* gene by direct complementation of a *C. albicans* histidine auxotroph using an improved double-ARS shuttle vector. *Gene* **165**, 115–120.
11. Gaur, N. A., Manoharlal, R., Saini, P., Prasad, T., Mukhopadhyay, G., Hoefer, M., Morschhauser, J., and Prasad, R. (2005) Expression of the *CDR1* efflux pump in clinical *Candida albicans* isolates is controlled by a negative regulatory element. *Biochem. Biophys. Res. Commun.* **332**, 206–214.
12. Li, D., Gurkovska, V., Sheridan, M., Calderone, R., and Chauhan, N. (2004) Studies on the regulation of the two-component histidine kinase gene *CHK1* in *Candida albicans* using the heterologous lacZ reporter gene. *Microbiology* **150**, 3305–3313.
13. El Barkani, A., Haynes, K., Mosch, H., Frosch, M., and Muhlschlegel, F. A. (2000) *Candida glabrata* shuttle vectors suitable for translational fusions to *lacZ* and use of beta-galactosidase as a reporter of gene expression. *Gene* **246**, 151–155.
14. Srikantha, T., Chandrasekhar, A., and Soll, D. R. (1995) Functional analysis of the promoter of the phase-specific *WH11* gene of *Candida albicans*. *Mol. Cell. Biol.* **15**, 1797–1805.
15. Staab, J. F., Bahn, Y. S., and Sundstrom, P. (2003) Integrative, multifunctional plasmids for hypha-specific or constitutive expression of green fluorescent protein in *Candida albicans*. *Microbiology* **149**, 2977–2986.
16. Barelle, C. J., Manson, C. L., MacCallum, D. M., Odds, F. C., Gow, N. A., and Brown, A. J. (2004) GFP as a quantitative reporter of gene

regulation in *Candida albicans. Yeast* **21**, 333–340.
17. Barelle, C. J., Priest, C. L., Maccallum, D. M., Gow, N. A., Odds, F. C., and Brown, A. J. (2006) Niche-specific regulation of central metabolic pathways in a fungal pathogen. *Cell. Microbiol.* **8**, 961–971.
18. Mao, Y., Zhang, Z., and Wong, B. (2003) Use of green fluorescent protein fusions to analyse the N- and C-terminal signal peptides of GPI-anchored cell wall proteins in *Candida albicans. Mol. Microbiol.* **50**, 1617–1628.
19. Horiguchi, H., Yurimoto, H., Goh, T., Nakagawa, T., Kato, N., and Sakai, Y. (2001) Peroxisomal catalase in the methylotrophic yeast *Candida boidinii*: transport efficiency and metabolic significance. *J. Bacteriol.* **183**, 6372–6383.
20. Eiden-Plach, A., Zagorc, T., Heintel, T., Carius, Y., Breinig, F., and Schmitt, M. J. (2004) Viral preprotoxin signal sequence allows efficient secretion of green fluorescent protein by *Candida glabrata*, *Pichia pastoris*, *Saccharomyces cerevisiae*, and *Schizosaccharomyces pombe. Appl. Environ. Microbiol.* **70**, 961–966.
21. Miyazaki, T., Tsai, H. F., and Bennett, J. E. (2006) Kre29p is a novel nuclear protein involved in DNA repair and mitotic fidelity in *Candida glabrata. Curr. Genet.* **50**, 11–22.
22. Staib, P., Moran, G. P., Sullivan, D. J., Coleman, D. C., and Morschhauser, J. (2001) Isogenic strain construction and gene targeting in *Candida dubliniensis. J. Bacteriol.* **183**, 2859–2865.
23. Palmer, G. E., Johnson, K. J., Ghosh, S., and Sturtevant, J. (2004) Mutant alleles of the essential 14-3-3 gene in *Candida albicans* distinguish between growth and filamentation. *Microbiology* **150**, 1911–1924.
24. Reynolds, A., Lundblad, V., Dorris, D., and Keaveney, M. (2000) Yeast vectors and assays for expression of cloned genes. In: *Current Protocols in Molecular Biology*, (Ausubel, F. M., Brent, B., Kingston, R. E., Moore, D. D., Seldman, J. G., Smith, J. A., and Struhl, K., eds.), John Wiley & Sons, Edison, NJ, pp. 13.61–13.66.
25. Brand, A., MacCallum, D. M., Brown, A. J. P., Gow, N. A. R., and Odds, F. C. (2004) Ectopic expression of *URA3* can influence the virulence phenotypes and proteome of *Candida albicans* but can be overcome by targeted reintegration of *URA3* at the *RPS10* locus. *Eukaryotic Cell* **3**, 900–909.
26. Strauss, A., Michel, S., and Morschhauser, J. (2001) Analysis of phase-specific gene expression at the single-cell level in the white-opaque switching system of *Candida albicans. J. Bacteriol.* **183**, 3761–3769.
27. Theiss, S., Kohler, G. A., Kretschmar, M., Nichterlein, T., and Hacker, J. (2002). New molecular methods to study gene functions in *Candida* infections. *Mycoses* **45**, 345–350.
28. Wirsching, S., Michel, S., Kohler, G., and Morschhauser, J. (2000) Activation of the multiple drug resistance gene *MDR1* in fluconazole-resistant, clinical *Candida albicans* strains is caused by mutations in a trans-regulatory factor. *J. Bacteriol.* **182**, 400–404.

Chapter 16

Genetic Transformation of *Candida albicans*

Ana M. Ramon and William A. Fonzi

Abstract

Genetic transformation is the primary method of genetic manipulation of *Candida albicans*. The lack of a complete sexual cycle prevents application of classical genetic analyses. However, transformation permits introduction into the genome of a wide variety of defined mutations including deletions, insertions, and fusions. Although several methods of transformation are available, the lithium-cation-induced transformation method described here is the most commonly used.

Key words: Transformation, cation-induced, lithium acetate, PEG (polyethylene glycol), spheroplasts, electroporation.

1. Introduction

Genetic analysis of *Candida albicans* is constrained by the absence of a complete sexual cycle. Because of this limitation, transformation is fundamental to genetic manipulation of this fungus. The seminal work of Kurtz et al. *(1)* demonstrated that spheroplasts of *C. albicans* could be transformed with exogenous DNA, which recombined with the homologous locus of the genome, and Kelly et al. *(2)* applied this approach to effect a one-step gene disruption. Since these early studies, a number of methodological improvements have enhanced the facility of transformation methods and many useful genetic transformation tools have been developed including recyclable auxotrophic and dominant-selectable drug markers and *(3–7)* PCR-amplifiable cassettes for gene deletion, epitope tagging, reporter-gene fusion, and promoter swapping *(8–11)*.

The spheroplast method used in the initial transformation studies *(1)*, though effective, was laborious to perform and the potential for spheroplast fusion and associated ploidy changes further diminished its desirability. This led to the development of more facile methods of transformation – cation-induced transformation *(12)* and electroporation *(13)*. Because lithium-cation-induced transformation is simple to perform and requires no specialized equipment, it is routinely used for transformation of *C. albicans*. The following is a general protocol based primarily on that of Geitz et al. *(14)* and Sanglard et al. *(12)*. Because transformation efficiency is influenced by many factors, such as genetic background of the test strain and structure of the transforming DNA, this procedure can serve as a starting point for optimization to suit specialized experimental situations.

2. Materials

1. YPD medium *(15)*: 2% glucose, 2% bacto peptone, 1% yeast extract. Sterilized by autoclaving.
2. 10X TE solution: 100 mM Tris–HCl, 10 mM EDTA, adjusted to pH 7.5. The solution is sterilized by autoclaving or filtration.
3. 10X LiAc solution: 1 M lithium acetate adjusted to pH 7.5 with acetic acid and filter sterilized (*see* **Note 1**).
4. 1X TE-LiAc is prepared fresh by mixing the appropriate volumes of 10X TE, 10X LiAc, and sterile dd H_2O.
5. PEG stock solution 50% (w/v) is prepared by dissolving 50 g of polyethylene glycol, average molecular weight 3000–4000 Da, in dd H_2O and adjusting the volume to 100 mL total. The solution may be filter-sterilized or autoclaved, but the viscosity of the solution makes it slow to filter.
6. PEG 40% is prepared fresh by mixing the appropriate volumes of 50% PEG, 10X TE, and 10X LiAc. For 10 transformations, mix 2400 µL of 50% PEG with 300 µL of 10X TE and 300 µL of 10X LiAc.
7. Single-stranded carrier DNA is prepared by dissolving 100 mg of DNA in 10 mL of 1X TE. Calf thymus, herring sperm, or salmon sperm DNA can be used. Once dissolved, the DNA is partially sheared by several passages through a pipet. The solution is dispensed in aliquots and stored at −20°C. Prior to use, an aliquot is placed in a boiling water bath for 5–10 min to denature the DNA and then quickly cooled on ice. A single aliquot can be used several times.

8. Transforming DNA is prepared according to the experimental design, either by PCR amplification or purification and restriction digestion of plasmid DNA.

9. Selection plates appropriate to the selection marker are prepared several days in advance and allowed to dry at room temperature.

3. Methods

1. The strain to be transformed is streaked from a frozen stock to an agar plate of YPD medium (*see* **Note 2**) and incubated 24–48 h at 30°C.

2. A single colony from the plate is inoculated into 10 mL of YPD medium in a 125 mL Erlenmeyer flask and incubated at 30°C on a shaker with good aeration for 16–24 h.

3. The cell density of the culture is determined by hemocytometer count or optical density measurement, and an aliquot is added to 50 mL of fresh YPD medium in a 250 mL Erlenmeyer flask to achieve a cell density of 2×10^6 cells/mL (approximately a 1:100 dilution). This culture is incubated at 30°C with aeration until a density of $1-2 \times 10^7$ cells/mL is reached, typically about 4 h (*see* **Note 3**).

4. The cells are harvested by centrifugation for 5 min at $5000 \times g$ at room temperature. The cell pellet is washed once with 10 mL of sterile distilled water, suspended in 1.0 mL of sterile distilled H_2O, and transferred to a 1.5 mL microcentrifuge tube.

5. The sample is centrifuged briefly to pellet the cells. The cell pellet is washed with 1.0 mL of sterile 1X TE-LiAc, suspended at a cell density of 2×10^9 cells/mL in 1X TE-LiAc, and placed on ice while the transforming DNA solution is prepared.

6. Transforming DNA solution is prepared by mixing 5 µL of single-stranded carrier DNA with 5 µL of transforming DNA (1–10 µg) in a 1.5 mL microcentrifuge tube.

7. A 50 µL aliquot of cells is added to the DNA solution and mixed by repeated pipetting.

8. PEG 40%, 300 µL, is added and mixed by repeated pipetting and/or vortexing (*see* **Note 4**).

9. The suspension is incubated for 30 min at 30°C, either stationary or with gentle mixing (*see* **Note 5**).

10. The suspension is placed in a 42°C water bath for 15 min to heat shock the cells (*see* **Note 6**).
11. The cells are collected by centrifugation in a microcentrifuge at room temperature for 30 s at 4000 × *g*.
12. Suspend the cell pellet in 300 µL of sterile distilled water and gently spread 50–150 µL on plates of the appropriate selection medium (*see* **Note 7**).
13. Incubate the plates at 30°C for 3–4 days.

4. Notes

1. At least one report has indicated the source of lithium acetate to be important for successful transformation *(16)*.
2. YPD is a rich medium that supports growth of most auxotrophs, but does not fully support growth of Ura3⁻ strains. If such strains are used, the medium should be supplemented with 25–100 µg/mL uridine. Uracil cannot be used in place of uridine, as *C. albicans* does not use uracil. Although YPD supports the growth of adenine auxotrophs, it has been reported that adenine supplementation of YPD enhances transformation of *Saccharomyces cerevisiae* Ade⁻ strains *(17)*. The effect on *C. albicans* transformation has not been reported.
3. Culture of 50 mL provides enough cells for 10 transformations, and the culture volume may be scaled up or down according to experimental needs. Logarithmically growing cells transform best, and transformation efficiency declines as cells approach stationary phase.
4. Some protocols stipulate addition of PEG solution prior to addition of DNA *(12)*, and the order of reagent addition may influence the efficiency of transformation *(18)*. The most effective protocol for *S. cerevisiae* combines all reagents prior to addition of the cells *(18)*.
5. Prolonged incubation of the transformation mix, as long as overnight, is reported to increase transformation efficiency four- to fivefold *(19)*.
6. Incubation at 44°C instead of 42°C can enhance transformation efficiency up to twofold *(19)*. The duration of heat shock also influences transformation *(18–20)*.
7. The amount of cells that need be plated depends on the efficiency of the transformation. This is influenced not only by the transformation protocol but also by a number of intrinsic factors. The frequency of integration can vary with the

expression level of the target locus *(21)*, the allele at a particular locus *(22)*, the length of homologous DNA flanking the selection marker *(23)*, and the genetic background of the strain *(18)*. As a practical matter, the entire suspension of cells should be plated to maximize recovery of low-frequency events. However, spreading an excessive number of cells on a single plate can limit the outgrowth of transformed cells or compromise selection when growth inhibitors are used. Consequently, the cells should be spread over several plates.

References

1. Kurtz, M. B., Cortelyou, M. W., and Kirsch, D. R. (1986) Integrative transformation of *Candida albicans*, using a cloned *Candida ADE2* gene. *Mol. Cell. Biol.* **6**, 142–149.
2. Kelly, R., Miller, S. M., and Kurtz, M. B. (1988) One-step gene disruption by cotransformation to isolate double auxotrophs in *Candida albicans*. *Mol. Gen. Genet.* **214**, 24–31.
3. Morschhauser, J., Michel, S., and Staib, P. (1999) Sequential gene disruption in *Candida albicans* by FLP-mediated site-specific recombination. *Mol. Microbiol.* **32**, 547–556.
4. Reuss, O., Vik, A., Kolter, R., and Morschhauser, J. (2004) The SAT1 flipper, an optimized tool for gene disruption in Candida albicans. *Gene* **341**, 119–127.
5. Wirsching, S., Michel, S., and Morschhauser, J. (2000) Targeted gene disruption in Candida albicans wild-type strains: the role of the MDR1 gene in fluconazole resistance of clinical Candida albicans isolates. *Mol. Microbiol.* **36**, 856–865.
6. Fonzi, W. A., and Irwin, M. Y. (1993) Isogenic strain construction and gene mapping in *Candida albicans*. *Genetics* **134**, 717–728.
7. Dennison, P. M., Ramsdale, M., Manson, C. L., and Brown, A. J. (2005) Gene disruption in *Candida albicans* using a synthetic, codon-optimised Cre-loxP system. *Fungal Genet. Biol.* **42**, 737–748.
8. Gerami-Nejad, M., Berman, J., and Gale, C. A. (2001) Cassettes for PCR-mediated construction of green, yellow, and cyan fluorescent protein fusions in *Candida albicans*. *Yeast* **18**, 859–864.
9. Gerami-Nejad, M., Hausauer, D., McClellan, M., Berman, J., and Gale, C. (2004) Cassettes for the PCR-mediated construction of regulatable alleles in *Candida albicans*. *Yeast* **21**, 429–436.
10. Wilson, R. B., Davis, D., and Mitchell, A. P. (1999) Rapid hypothesis testing with *Candida albicans* through gene disruption with short homology regions. *J. Bacteriol.* **181**, 1868–1874.
11. Schaub, Y., Dunkler, A., Walther, A., and Wendland, J. (2006) New pFA-cassettes for PCR-based gene manipulation in *Candida albicans*. *J. Basic Microbiol.* **46**, 416–429.
12. Sanglard, D., Ischer, F., Monod, M., and Bille, J. (1996) Susceptibilities of *Candida albicans* multidrug transporter mutants to various antifungal agents and other metabolic inhibitors. *Antimicrob. Agents Chemother.* **40**, 2300–2305.
13. De Backer, M. D., Maes, D., Vandoninck, S., Logghe, M., Contreras, R., and Luyten, W. H. (1999) Transformation of *Candida albicans* by electroporation. *Yeast* **15**, 1609–1618.
14. Gietz, D., St. Jean, A., Woods, R. A., and Schiestl, R. H. (1992) Improved method for high efficiency transformation of intact yeast cells. *Nucl. Acids Res.* **20**, 1425.
15. Sherman, F., Fink, G. R., and Hicks, J. B. (1986) *Methods in Yeast Genetics*. Cold Spring Harbor Laboratories, Cold Spring Harbor, NY.
16. Hull, C. M., and Johnson, A. D. (1999) Identification of a mating type-like locus in the asexual pathogenic yeast *Candida albicans*. *Science* **285**, 1271–1275.
17. Schiestl, R. H., and Gietz, R. D. (1989) High efficiency transformation of intact yeast cells using single stranded nucleic acids as a carrier. *Curr. Genet.* **16**, 339–346.

18. Gietz, R. D., and Woods, R. A. (2002) Transformation of yeast by lithium acetate/single-stranded carrier DNA/polyethylene glycol method. *Methods Enzymol.* **350**, 87–96.
19. Walther, A., and Wendland, J. (2003) An improved transformation protocol for the human fungal pathogen *Candida albicans*. *Curr. Genet.* **42**, 339–343.
20. Braun, B. R., and Johnson, A. D. (1997) Control of filament formation in *Candida albicans* by the transcriptional repressor *TUP1*. *Science* **277**, 105–109.
21. Srikantha, T., Morrow, B., Schröppel, K., and Soll, D. R. (1995) The frequency of integrative transformation at phase-specific genes of *Candida albicans* correlates with their transcriptional state. *Mol. Gen. Genet.* **246**, 342–352.
22. Yesland, K., and Fonzi, W. A. (2000) Allele-specific gene targeting in *Candida albicans* results from heterology between alleles. *Microbiology* **146**, 2097–2104.
23. Gola, S., Martin, R., Walther, A., Dunkler, A., and Wendland, J. (2003) New modules for PCR-based gene targeting in *Candida albicans*: rapid and efficient gene targeting using 100 bp of flanking homology region. *Yeast* **20**, 1339–1347.

Chapter 17

Large-Scale Gene Disruption Using the *UAU1* Cassette

Clarissa J. Nobile and Aaron P. Mitchell

Abstract

Candida albicans is a major fungal systemic pathogen in humans. Genetic manipulation of *C. albicans* is unwieldy. We report here a strategy that is useful and successful for large-scale genetic manipulation of *C. albicans* genes of interest: use of the *UAU1* cassette on a *Tn7* transposon. Streamlined yet admittedly flawed disruption techniques, such as the one described here, may prove vital to uncovering the genetic basis of fungal virulence.

Key words: *UAU1*, large-scale gene disruption, mutants.

1. Introduction

Until recently, *Candida albicans* was considered a genetically intractable organism for several reasons. First, its genome was not sequenced. Second, it is a diploid that lacks a complete sexual cycle, thus making null mutants requires two successive transformations. Lastly, *C. albicans* transformations are not very efficient. From the time that the first *C. albicans* gene was disrupted in 1993 *(1)* and up until November 2002, less than 150 mutants had been defined out of greater that 6000 open reading frames (ORFs) in the haploid complement of its genome. Now in this postgenomic era, we have easily accessible online databases containing the DNA sequences for our favorite *C. albicans* ORFs (see The *Candida* Genome Database at www.candidagenome.org; and the Pasteur Institute CandidaDB at http://genodb.pasteur.fr/cgi-bin/WebObjects/CandidaDB). The accessibility of the fully sequenced genome should aid in the first difficult part of a large-scale gene disruption procedure – deciding on a group of genes to disrupt.

Once a group of genes has been chosen, the hard part is the actual large-scale genetic manipulation. We describe here a procedure that we find useful and successful for the rapid production of homozygous null mutants through just a single transformation. This disruption procedure involves the use of the *UAU1* cassette (**Fig. 17.1**) *(2)*. The *UAU1* cassette consists of a functional

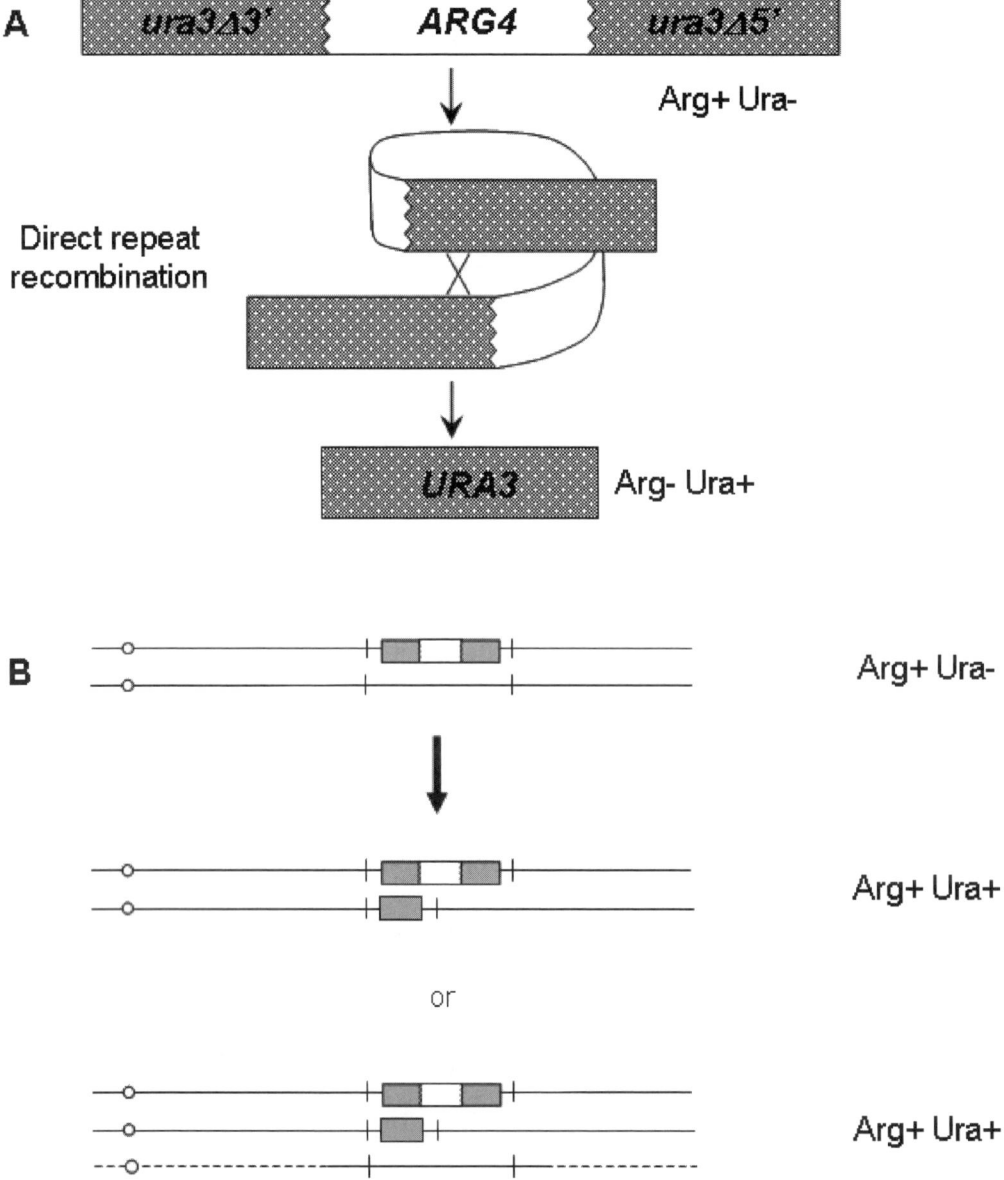

Fig. 17.1. Genetic properties of the *UAU1* cassette. (**A**) Conversion of *UAU1* to *URA3*. Recombination excises the *ARG4* gene and results in an Arg$^-$ Ura$^+$ phenotype. (**B**) Outcome of double-disruption selection with the *UAU1* cassette yields the intended homozygote disruption mutant and/or a triplication segregant.

C. albicans *ARG4* gene flanked by nonfunctional *URA3* deletion derivatives (*ura3Δ3'-ARG4-ura3Δ5'*), thus conferring an Arg⁺ Ura⁻ phenotype after transformation when the cassette has inserted into the first allele. At this point, the heterozygous *UAU1* insertion can become homozygous through a mitotic gene conversion event. Then, by selecting for the Arg⁺ Ura⁺ phenotype, we select for those segregants that have undergone a mitotic gene conversion, followed by a looping out of the *ARG4* gene between the *URA3* segments in one of the alleles (the *URA3* segments share approximately 500 bp of overlapping homology for recombination), thus yielding an intact *URA3* gene. Homozygous Arg⁺ Ura⁺ isolates can then be genotyped using colony polymerase chain reaction (PCR) for the absence of a wild-type allele, and the presence of the insertion allele. Typically, after screening all of the Arg⁺ Ura⁺ isolates by PCR, two types of genotypes are observed. The first type consists of the intended homozygote disruption mutants, and the second consists of allelic triplications, where a copy of the wild-type allele is present in addition to the two disrupted alleles. It is unknown how this triplication event occurs, but one could fathom that it may occur via a translocation, a tandem duplication, or nondisjunction.

This triplication byproduct of using the *UAU1* cassette to disrupt genes has its advantages. Since deletion of essential genes, by definition, results in lethality, failure to obtain null deletion mutants of a gene has led to the assumption that said gene is essential. This may not always be the case; it is possible that the gene may not be essential, but simply difficult to delete due to such things as a highly repetitive sequence, or the presence of closely related genes. The *UAU1* cassette is an excellent test to identify potential essential genes. We deduce that a *UAU1* insertion in nonessential genes produces both homozygous mutants and triplication segregants, whereas in essential genes, the *UAU1* cassette yields only triplication segregants *(2)*. Since essential genes are putative drug targets, the *UAU1* cassette may prove beneficial in this area. Thus, the *UAU1* cassette is useful for both the rapid and systematic disruption of genes, as well as for the identification of essential genes.

2. Materials

2.1. Reagents for Preparation of C. albicans Genomic DNA

1. YPD: (10 g yeast extract, 20 g peptone, 20 g glucose, to 1000 mL with dH$_2$0) liquid media. Supplement YPD medium with 80 μg/μL uridine if using a Ura⁻ *C. albicans* strain (*see* **Note 1**).
2. TENTS: (1% SDS, 2% Triton X-100, 0.1 M NaCl in 1XTE). Filter sterilize.

3. Acid washed beads (sterilize by autoclaving).
4. Phenol/chloroform/isoamyl alcohol (25:24:1, v/v).
5. Ethanol.
6. 1 × Tris-EDTA (1 × TE), pH 7.4
7. 10 mg/mL RNase.
8. 10 M NH_4OAc.

2.2. PCR Reagents

1. 1 × Tris-EDTA (1 × TE), pH 7.4.
2. 10X PCR buffer (100 mM Tris–HCl, pH 8.3, 500 mM KCl, 15 mM $MgCl_2$, 0.01% gelatin).
3. 25 mM $MgCl_2$.
4. 5 mM dNTPs.
5. Taq DNA polymerase.
6. 10 µM forward ORF primer.
7. 10 µM reverse ORF primer.

2.3. Ligation Reagents

1. pGEMT-Easy vector (Promega).
2. T4 DNA ligase (Promega).
3. 2 × Ligation buffer (Promega).

2.4. Materials for E. coli Transformation, Plasmid Extraction/Purification, and Cloning Confirmation

1. Chemically competent *Escherichia coli*.
2. LB+Amp (100 µg/mL) + X-Gal (40 mg/mL) plates.
3. LB+Amp (100 µg/mL) liquid media.
4. Montage Plasmid Miniprep$_{96}$ kit (Millipore) (*see* **Note 2**).
5. Restriction enzyme NotI (New England Biolabs) (*see* **Note 3**).
6. 10 × NE Buffer 3 (New England Biolabs).
7. 100 × BSA (New England Biolabs).

2.5. Reagents for GPS Transposon Mutagenesis

1. Donor plasmid pDDB166 (**Fig. 17.2**).
2. Target plasmid pGEMT-Easy-ORF.
3. 10 × GPS buffer (New England Biolabs).
4. TnsABC* Transposase (New England Biolabs).
5. Start solution (New England Biolabs).
6. PI-SceI (New England Biolabs).
7. 10 × PI-SceI buffer (New England Biolabs).
8. 100 × BSA (New England Biolabs).

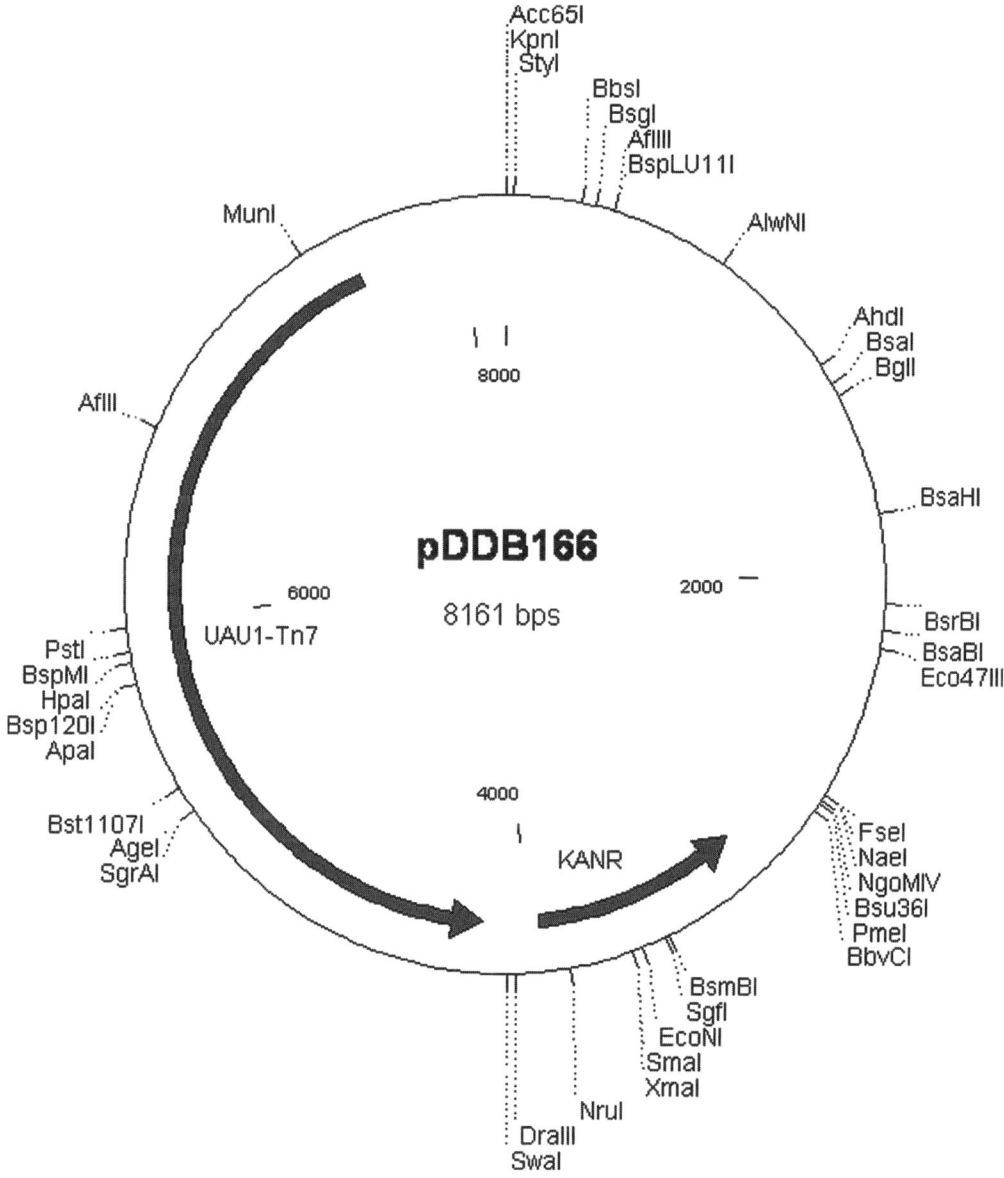

Fig. 17.2. Donor plasmid pDDB166 containing the *UAU1* cassette on a *Tn7* transposon. Plasmid pDDB166 confers kanamycin resistance.

2.6. Reagents for Transformation of Mutagenesis Reaction into E. coli, Extraction of Mutagenized Plasmid DNA, and Confirmation of Mutagenesis into ORF of Interest

1. Chemically competent *E. coli*.
2. LB + Amp + Kan (100 μg/mL and 50 μg/mL, respectively) plates.
3. LB liquid media.
4. LB + Amp + Kan (100 μg/mL and 50 μg/mL, respectively) liquid media.
5. Montage Plasmid Miniprep$_{96}$ kit (Millipore) (*see* **Note 2**).
6. Restriction enzyme NotI (New England Biolabs) (*see* **Note 4**).

7. 10X NE buffer 3 (New England Biolabs).
8. 100X BSA (New England Biolabs).
9. Restriction enzyme PmeI (New England Biolabs).
10. 10X NE buffer 2 (New England Biolabs).

2.7. Reagents for Digestion of Plasmids for Transformation and Transformation into C. albicans

1. *C. albicans* strain BWP17 *(3)*.
2. YPD + Uri (80 µg/µL) plates.
3. YPD + Uri (80 µg/µL) liquid media.
4. Extracted plasmid DNA containing *Tn7-UAU1* in middle of ORF.
5. Restriction enzyme NotI (New England Biolabs) (*see* **Note 4**).
6. 10X NE buffer 3 (New England Biolabs).
7. 100X BSA (New England Biolabs).
8. LATE: 0.1 M LiOAc in 1X TE buffer.
9. Calf thymus DNA (9.1 mg/mL) (Sigma).
10. PLATE: 8 mL 50% PEG + 1 mL 10X TE + 1 mL 1 M LiOAc.
11. YPD plates.
12. SC-Arg plates.
13. SC-Arg-Ura plates.

2.8. Reagents for Confirmation of Homozygous Insertion Mutants

1. 10 µM Arg4-detect primer (sequence GGAATTGATCAAT-TATCTTTTGAAC) (*see* **Note 5**).
2. 10 µM forward ORF primer.
3. 10 µM reverse ORF primer.
4. 10X PCR buffer (100 mM Tris-HCl, pH 8.3, 500 mM KCl, 15 mM $MgCl_2$, 0.01% gelatin).
5. 25 mM $MgCl_2$.
6. 5 mM dNTPs.
7. Taq DNA polymerase.

3. Methods

The methods to follow describe in detail a large-scale *C. albicans* disruption procedure in a step-by-step format. We begin with a description of the preparation of *C. albicans* genomic DNA for use in a PCR reaction to amplify your ORFs of interest. We follow with in-depth instructions on large-scale PCR protocols and cloning of the ORFs into a suitable plasmid vector. We then describe a transposon mutagenesis protocol using the *UAU1* cassette on a *Tn7* transposon into the plasmid vector containing the ORFs of interest, followed by the final selection of *C. albicans* homozygous disruption mutants.

3.1. Preparation of C. albicans Genomic DNA

1. Inoculate a 5-mL culture of YPD or YPD + Uri (*see* **Note 1**) liquid media with *C. albicans* reference strain (*see* **Note 6**). Grow the culture overnight with agitation at 30°C.
2. Spin down the entire 5 mL culture at low speed and aspirate supernatant.
3. Add 500 μL TENTS and resuspend.
4. Transfer the resuspended mixture to a fresh eppendorf tube containing 200 μL of sterile acid washed beads.
5. Add 500 μL phenol/choloroform/isoamylalcohol.
6. Vortex for 2 min.
7. Spin down tubes at 14,000 rpm at 4°C for 10 min.
8. Transfer aqueous phase to a new eppendorf tube.
9. Add 1 mL 100% ethanol.
10. Place at –20°C for at least 1 h.
11. Spin down tubes at 14,000 rpm at 4°C for 15 min.
12. Aspirate supernatant, and resuspend in 200 μL 1 × TE.
13. Add 1 μL 10 mg/mL RNase.
14. Place at room temperature for 30 min.
15. Add 40 μL 10 M NH_4OAc.
16. Add 500 μL 100% ethanol and mix by inversion.
17. Place at –20°C for 30 min.
18. Spin down tubes at 14,000 rpm for 5 min. Decant supernatant.
19. Add 1 mL 70% Ethanol to pellet and immediately decant.
20. Place tubes open in a speed vacuum for 5–10 min or until dry.
21. Resuspend gently in 50 μL 1 × TE (*see* **Note 7**).

3.2. PCR of ORFs of Interest

1. Primers are designed for amplification of ORFs of interest from the start codon to the stop codon using 20 bp oligonucleotides. Primers are diluted to 100 μM in 1 × TE, and further diluted to a working stock of 10 μM in dH_2O.
2. PCR is performed off of a diluted genomic DNA template (prepared in **Section 3.1**) isolated from a *C. albicans* reference

strain containing the ORFs of interest for amplification (*see* **Note 6**).

3. Dilute genomic DNA template for amplification by PCR. If genomic DNA was isolated as in **Section 3.1**, a 1:1000 dilution in dH_2O is recommended for the PCR reaction.

4. An example of a typical large-scale PCR reaction for one sample is listed below (*see* **Note 8**). Mix well.

10X PCR buffer	5 µL
25 mM $MgCl_2$	1 µL
5 mM dNTPs	2 µL
Genomic DNA template (1/1000)	1 µL
10 µM forward ORF primer	2 µL
10 µM reverse ORF primer	2 µL
Taq DNA polymerase	0.2 µL
dH_2O	36.8 µL
Total volume	50 µL

5. An example of a typical large-scale PCR program for amplification of most *C. albicans* ORFs is listed below (*see* **Note 9**).

Step 1	94°C for 3 min
Step 2	94°C for 1 min
Step 3	50°C for 1 min
Step 4	72°C for 4 min
Step 5	35 times to **Step 2**
Step 6	72°C for 10 min
Step 7	4°C/End

6. Run 5 µL of PCR product on an agarose gel containing ethidium bromide. Typically, a single band of the correct size for the ORF is present. No purification of the PCR sample is necessary for the ligation reaction to follow (*see* **Note 10**).

3.3. Ligation of PCR Products into Plasmid Vector pGEMT-Easy

1. Ligate PCR product into pGEMT-Easy as detailed below.

2 × ligation buffer	7.5 µL
T4 DNA ligase	1 µL
pGEMT-Easy vector	1 µL
PCR product insert	3 µL
dH$_2$0	2.5 µL
Total Volume	15 µL

2. Mix gently.
3. Allow reaction to ligate for 2–4 h at room temperature or overnight at 4°C.

3.4. Transformation of Ligation Reaction into E. coli, Extraction of Plasmid DNA, and Cloning Confirmation

1. Thaw chemically competent *E. coli* on ice.
2. Add entire 15 µL ligation reaction to appropriate volume of thawed chemically competent *E. coli*.
3. Allow mixture to rest on ice for 10 min.
4. Heat shock mixture at 42°C for 45 s.
5. Place on ice for an additional 3 min.
6. Plate directly on LB + Amp + X-Gal plates for selection of the ORF inserted into the pGEMT-Easy vector. Grow overnight at 37°C.
7. Select eight white colonies and inoculate separately into 2 mL LB + Amp liquid media.
8. Grow overnight at 37°C with agitation.
9. Spin down cultures and extract plasmid DNA using the Montage Plasmid Miniprep$_{96}$ kit (*see* **Note 2**).
10. Digest a portion of extracted plasmid DNA in order to determine if the ORF of interest has inserted correctly into the pGEMT-Easy vector. An example digestion is listed below using the restriction enzyme NotI (*see* **Note 3**).

Extracted plasmid DNA	3 µL
10 × NE Buffer 3	2 µL
100 × BSA	0.2 µL
NotI	0.2 µL
dH$_2$O	14.6 µL
Total volume	20 µL

11. Mix gently.
12. Allow the reaction to digest for 2 h at 37°C.
13. Run 10 μL of digestion reaction on an agarose gel containing ethidium bromide. Using NotI releases an approximately 3-kb band for the pGEMT-Easy vector backbone. The gel should thus resolve an approximately 3-kb band representing the vector, as well as a band of the expected size for the inserted ORF of interest (*see* **Note 11**).

3.5. GPS Transposon Mutagenesis

The following steps require the use of the plasmid pDDB166 as the donor plasmid containing the *UAU1* cassette on a *Tn7* transposon (**Fig. 17.2**). Plasmid pDDB166 was constructed by destroying the ScaI site of pAED98 *(3)* so that it no longer confers resistance to ampicillin, but retains resistance to kanamycin. The target plasmid for the mutagenesis reaction is the pGEMT-Easy plasmid containing the inserted ORF of interest (or pGEMT-Easy-ORF) for disruption. Donor plasmid pDDB166 is kanamycin resistant and target plasmid pGEMT-Easy-ORF is ampicillin resistant. When mutagenesis into the target plasmid is successful, clones will be both ampicillin- and kanamycin resistant. The mutagenesis reaction described below is modified from the GPS-Mutagenesis System instruction manual (New England Biolabs).

1. Dilute donor plasmid pDDB166 and target plasmid pGEMT-Easy-ORF to a concentration of 20 ng/μL (*see* **Note 12**).
2. Add the following reagents for the GPS mutagenesis reaction:

10 × GPS buffer	2 μL
Donor pDDB166 (20 ng/μL)	1 μL
Target pGEMT-Easy-ORF (20 ng/μL)	4 μL
dH$_2$0	11 μL
Total volume	18 μL

3. Mix gently.
4. Add 1 μL TnsABC* Transposase and mix gently.
5. Place the reaction at 37°C for 15 min.

6. Add 1 μL Start Solution, and mix gently.
7. Place the reaction at 37°C for 1 h.
8. Heat inactivate the transposase at 75°C for 10 min.
9. To destroy the pDDB166 donor backbone, add the following reagents to the reaction:

10X PI-SceI buffer	5 μL
100X BSA	0.5 μL
PI-SceI	6 μL
dH$_2$0	18.5 μL
Total volume	30 μL

10. Place the reaction at 37°C for 2–3 h.
11. Place the reaction at 75°C for 10 min.

3.6. Transformation of Mutagenesis Reaction into E. coli, Extraction of Mutagenized Plasmid DNA, and Confirmation of Mutagenesis into ORF of Interest

1. Thaw chemically competent *E. coli* on ice.
2. Add 10 μL of undiluted mutagenesis reaction to appropriate volume of thawed chemically competent *E. coli*.
3. Allow mixture to rest on ice for 10 min.
4. Heat shock mixture at 42°C for 45 s.
5. Place on ice for an additional 5 min.
6. Add 1 mL LB, and incubate for 1 h at 37°C with agitation for outgrowth.
7. Spin down cells at low speed for 1 min, decant supernatant, and resuspend cell pellet in 100 μL dH$_2$O.
8. Plate on LB + Amp + Kan plates. Grow overnight at 37°C.
9. Select eight white colonies, and inoculate separately into 2 mL LB + Amp + Kan liquid media.
10. Grow overnight at 37°C with agitation.
11. Spin down cultures and extract plasmid DNA using the Montage Plasmid Miniprep$_{96}$ kit (*see* **Note 2**).
12. Digest a portion of the extracted plasmid DNA in order to determine if the *Tn7-UAU1* cassette has inserted into the ORF of interest in the target plasmid.

An example digestion is listed below using the restriction enzyme NotI (*see* **Note 4**).

Extracted plasmid DNA	3 µL
10X NE Buffer 3	2 µL
100X BSA	0.2 µL
NotI	0.2 µL
dH$_2$O	14.6 µL
Total volume	20 µL

13. Mix gently.
14. Allow the reaction to digest for 2 h at 37°C.
15. Run 10 µL of digestion reaction on an agarose gel containing ethidium bromide.
16. Two scenarios are detectable on a gel when the *Tn7-UAU1* cassette is inserted into the pGEMT-Easy-ORF target plasmid, and digested with NotI (*see* **Note 13**). The first possibility is insertion of the *Tn7-UAU1* cassette into the pGEMT-Easy backbone of the pGEMT-Easy-ORF plasmid. The second possibility is insertion of the *Tn7-UAU1* cassette into the ORF of the pGEMT-Easy-ORF plasmid. The latter is the desired scenario. **Figure 17.3** presents an example mutagenesis of an ORF of size 1800 bp. Clones 1, 2, 3, 7, and 8 are examples of the desired scenario where the *Tn7-UAU1* cassette has inserted successfully into the 1800 bp ORF (**Fig. 17.3**, *upper band* for clones 1–3, and 7–8), leaving behind the

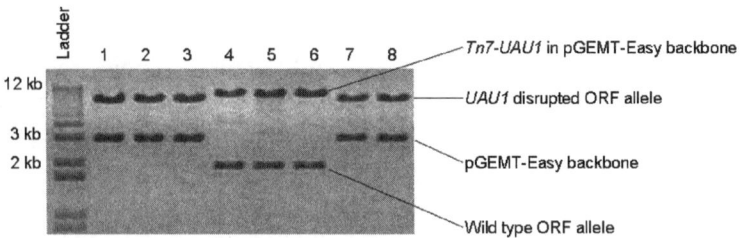

Fig. 17.3. Mutagenesis of an ORF of size 1800 bp. Clones 1, 2, 3, 7, and 8 are examples of where the *Tn7-UAU1* cassette has inserted successfully into the 1800 bp ORF, leaving behind the pGEMT-Easy backbone of approximately 3 kb upon NotI digestion. Clones 4, 5, and 6 are examples of where the *Tn7-UAU1* cassette has inserted into the pGEMT-Easy backbone of the pGEMT-Easy-ORF plasmid; not into the ORF, leaving behind the ORF of 1800 bp upon NotI digestion.

pGEMT-Easy backbone of approximately 3 kb (**Fig. 17.3**, lower approximately 3 kb band for clones 1–3, and 7–8) upon NotI digestion. Clones 4, 5, and 6 are examples of where the *Tn7-UAU1* cassette has inserted into the pGEMT-Easy backbone of the pGEMT-Easy-ORF plasmid, not into the ORF (**Fig. 17.3**, *upper band* for clones 4–6), leaving behind the ORF of 1800 bp (**Fig. 17.3**, lower 1800 bp band for clones 4–6) upon NotI digestion.

17. Digest a portion of the desired scenario extracted plasmid DNA, where the *Tn7-UAU1* cassette has inserted successfully into the ORF of interest, in order to approximate where the *Tn7-UAU1* cassette has inserted into the ORF. An example digestion is listed below using the restriction enzymes NotI and PmeI; a PmeI site is present on one end on the *Tn7-UAU1* cassette (**Fig. 17.2**).

Extracted plasmid DNA containing *Tn7-UAU1* in ORF	3 μL
10X NE Buffer 2	2 μL
100X BSA	0.2 μL
NotI	0.2 μL
PmeI	0.2 μL
dH$_2$O	14.4 μL
Total volume	20 μL

18. Mix gently.
19. Allow the reaction to digest for 2 h at 37°C.
20. Run 10 μL of digestion reaction on an agarose gel containing ethidium bromide.
21. Since the PmeI site is on one end of the *Tn7-UAU1* cassette, digestion with PmeI and NotI for the clones, where the *Tn7-UAU1* cassette has inserted into the ORF, will produce three detectable bands on an agarose gel (**Fig. 17.4**). We will focus on the smallest band because this band is the size from one end of the *Tn7-UAU1* cassette to the closest adjacent end of the ORF (*see* **Note 14**). **Figure 17.4** presents an example of how we confirmed where the *Tn7-UAU1* cassette inserted into the 1800 bp ORF from **Fig. 17.3** for clones 1, 2, 3, 7, and 8. It is evident from

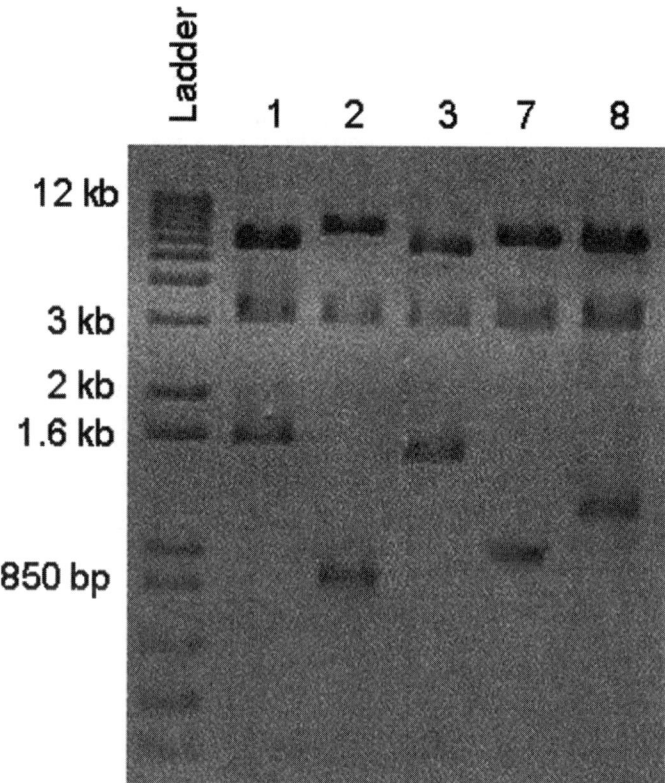

Fig. 17.4. Confirmation of location in which the *Tn7-UAU1* cassette has inserted into the 1800 bp ORF for clones 1, 2, 3, 7, and 8. Clones 2 and 7 contain the *Tn7-UAU1* cassette inserted approximately into the middle of the 1800 bp ORF, as the smallest bands for digested clones 2 and 7 are approximately 900 and 980 bp, respectively.

the mutagenesis reaction that the *Tn7-UAU1* cassette hopped into different regions of the ORF (**Fig. 17.4**). For the sake of simplicity, we will go further with clones 2 and 7 because the *Tn7-UAU1* cassette inserted approximately into the middle of the 1800 bp ORF as the smallest bands for digested clones 2 and 7 are approximately 900 and 980 bp, respectively (**Fig. 17.4**).

3.7. Digestion of Plasmids Containing the Tn7-UAU1 Cassette in the Middle of the ORF and Transformation into C. albicans

1. Streak out *C. albicans* strain BWP17 *(4)* for singles on a YPD + Uri plate. Grow at 30°C for 2 days.
2. Culture a single colony in 5 mL YPD + Uri liquid media. Grow overnight at 30°C with agitation.
3. The following day, determine the OD_{600} of the overnight culture.
4. Dilute the sample to $OD_{600} = 0.2$ in 50 mL YPD + Uri liquid media.
5. Incubate the diluted culture at 30°C until the culture reaches an OD_{600} of approximately 0.8 (*see* **Note 15**).

6. During this incubation period, digest 10 μL of each extracted plasmid containing the *Tn7-UAU1* cassette inserted into the middle of the ORF of interest with a restriction enzyme that releases the ORF containing the *Tn7-UAU1* cassette from the pGEMT-Easy vector backbone. An example digestion is listed below using the restriction enzyme NotI.
7. Mix gently.

Extracted plasmid DNA containing *Tn7-UAU1* in middle of ORF	10 μL
10X NE Buffer 3	2 μL
100X BSA	0.2 μL
NotI	0.5 μL
dH$_2$O	7.3 μL
Total volume	20 μL

8. Allow the reaction to digest at 37°C until the *C. albicans* culture is ready for transformation (approximately 3–4 h).
9. When the *C. albicans* culture has reached an OD$_{600}$ of approximately 0.8, pour the culture into a 50-mL conical tube and spin at low speed (approximately 5000 rpm) for 5 min.
10. Discard the supernatant and wash the cell pellet by gently resuspending it in 5 mL sterile dH$_2$O (do not vortex).
11. Spin at low speed for 5 min and discard the supernatant.
12. Resuspend the cell pellet in 500 μL of LATE.
13. To set up *C. albicans* transformation reactions, add for each transformation

LATE cell suspension	100 μL
Calf thymus DNA	10 μL
Digest from **Step 6**	20 μL

14. Mix gently.
15. Incubate for 30 min at 30°C.
16. Add 700 μL of freshly made PLATE and incubate overnight at 30°C (*see* **Note 16**).

17. Heat shock the cell mixture at 44°C for 20 min.
18. Spin cells down for 30 s at low speed and aspirate the supernatant.
19. Wash the cell pellet by resuspending in 1 mL YPD + Uri.
20. Spin cells down for 30 s at low speed and decant the supernatant.
21. Resuspend the cells gently in 100 μL of YPD + Uri and plate on Beefy SC-Arg plates.
22. Grow for 2 days at 30°C.
23. Pick 12 colonies from each transformation plate, streak for singles on SC-Arg plates, and grow for 2 days at 30°C. These transformants should be heterozygous mutants.
24. Patch a single colony from each of the 12 heterozygote mutants (streaked for singles) onto YPD plates (*see* **Note 17**). Grow for 2 days at 30°C.
25. Replica plate the YPD patches onto SC-Arg-Ura plates.
26. Grow for 4–5 days at 30°C. Some of these transformants should be homozygous mutants.
27. Pick one colony from each quarter and streak for singles on SC-Arg-Ura plates. Grow for 2 days at 30°C.

3.8. Confirmation of Homozygous Insertion Mutants

28. Colony PCR directly off of a single colony using three primers. An example of a typical large-scale colony PCR reaction for one sample is listed below.

10X PCR buffer	5 μL
25 mM $MgCl_2$	3 μL
5 mM dNTPs	2 μL
10 μM Arg4-detect primer	1 μL
10 μM forward ORF primer	1 μL
10 μM reverse ORF primer	1 μL
dH_2O	37 μL
Total volume	50 μL

29. Add a single colony to the PCR mixture.
30. Before adding the tubes to the PCR block, start the program and pause it at the first step at 94°C. When the block has

reached 94°C, add the tubes containing the PCR mixture with the colony. Allow the tubes to boil in the thermocycler (paused at 94°C) for 8 min.

31. Add 0.5 µL Taq DNA polymerase to each tube.
32. Unpause the program and let it run to completion. The colony PCR program is listed below.

Step 1	94°C for 2 min
Step 2	94°C for 45 s
Step 3	50°C for 45 s
Step 4	72°C for 4 min
Step 5	35 times to **Step 2**
Step 6	72°C for 12 min
Step 7	4°C/End

33. Run 10 µL of the colony PCR on an agarose gel containing ethidium bromide. **Figure 17.5** shows an example of a colony PCR from a mutagenesis of an ORF of size 1800 bp. The heterozygote mutant (*far left lane*) for the corresponding

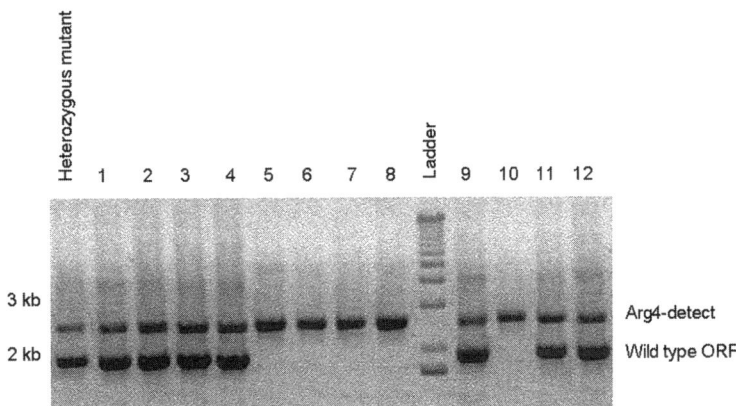

Fig. 17.5. Colony PCR from a mutagenesis of an ORF of size 1800 bp. The heterozygote mutant (*far left lane*) for the corresponding putative homozygote mutants (colonies 1–12) contains the Arg4-detect band (*upper band*) and the lower 1800 bp band corresponding to the size of the wild-type ORF. PCR off of the putative homozygous mutant colonies 1–12 shows two different scenarios. PCR off of colonies 1–4, 9, 11, and 12 resemble the banding pattern of the heterozygote mutant with the Arg4-detect band (*upper band*) and the lower 1800 bp band corresponding to the size of the wild type ORF; these colonies are triplication derivatives. PCR off of colonies 5–8, and 10 contain just the Arg4-detect band (*upper band*); these colonies are homozygous mutants.

putative homozygote mutants (colonies 1–12) contains the Arg4-detect band (*upper band*) and the lower 1800-bp band corresponding to the size of the ORF (**Fig. 17.5**). PCR off of the putative homozygous mutant colonies 1–12 shows two different scenarios. PCR off of colonies 1–4, 9, 11, and 12 resembles the banding pattern of the heterozygote mutant with the Arg4-detect band (*upper band*) and the lower 1800-bp band corresponding to the size of the ORF (**Fig. 17.5**); these colonies are not homozygous mutants, but correspond to triplication derivatives. PCR off of colonies 5–8, and 10 contain just the Arg4-detect band (*upper band*) (**Fig. 17.5**); these colonies are homozygous mutants.

4. Notes

1. *C. albicans* strains that are Ura– (with the *URA3* gene disrupted) require supplementation with uridine at 80 mg/L because disruption of *URA3* blocks the ability to synthesize uridine. Supplementation with uracil, which is typical for *Saccharomyces cerevisiae* media recipes, is not adequate for growth: *C. albicans* may lack the ability to take uracil up efficiently or to employ the uracil salvage pathway.

2. Plasmid DNA can be extracted and purified using several methods or several commercially available kits depending on the number of samples. For a small number of samples, the standard alkaline lysis or boiling mini-preparation methods are acceptable. For a medium number of samples, the QIAprep Spin Miniprep kit (Qiagen) is ideal. For a large number of samples, we recommend the Montage Plasmid Miniprep$_{96}$ kit (Millipore) because it allows the rapid extraction and purification of plasmid DNA in 96-well plates for large-scale procedures.

3. Several restriction enzymes could be used to confirm the presence of the ORF insert into the pGEMT-Easy vector. We use NotI here because it is a rare cutter, and is not present in most *C. albicans* ORFs. Thus, it is most useful for a large-scale procedure. However, if NotI is present in your ORF of interest, several other convenient restriction enzymes are present in the pGEMT-Easy vector (see Promega pGEMT-Easy manual for details).

4. Several restriction enzymes could be used to confirm the presence of the *Tn7-UAU1* cassette into the ORF of interest in the target plasmid. If NotI is present in the ORF of interest, PvuII is another useful alternative.

5. Arg4-detect primer is a reverse primer on one end of *ARG4*.

6. We typically extract genomic DNA from reference strains BWP17 (Arg⁻ Ura⁻ His⁻) *(4)*, DAY185 (Arg⁺ Ura⁺ His⁺) *(5)*, or DAY286 (Arg⁺ Ura⁺ His⁻) *(3)*.

7. Genomic DNA is extremely delicate. To avoid shearing, genomic DNA should never be vortexed.

8. This is an example PCR reaction that works for the amplification of most *C. albicans* ORFs. Harder to amplify ORFs may require other additions.

9. This is a generalized PCR program that works for most *C. albicans* ORFs less than 4 kb in length. Larger ORFs may require longer extension times.

10. It is unnecessary to clean up the PCR sample for direct ligation into plasmid vector pGEMT-Easy as long as a single band of the correct size is present. In fact, we find that purification of the sample using the Qiagen PCR purification and/or gel extraction kits significantly reduces the DNA concentration of the sample as well as the ligation efficiency, and should only be used if multiple bands are present.

11. Certain ORFs are the same size or close in size to approximately 3 kb pGEMT-Easy vector backbone that is released when the ORF-containing plasmid is digested with NotI. If this is the case, other restriction enzymes should be considered for use instead of NotI that cut at different sites within the vector backbone, thus altering the size of the released vector backbone fragment.

12. The mutagenesis reaction is extremely sensitive to the ratio of donor to target plasmid concentration, and thus concentrations of plasmids should be exact.

13. The insertion of the *Tn7-UAU1* cassette into the ORF of interest versus the target plasmid backbone is dependent on the size of the ORF. The chance of the transposon hopping into the ORF is based on probability. Larger ORFs have a higher likelihood of being hopped into versus smaller ORFs. For smaller ORFs, it is not a bad idea to screen more clones.

14. We chose to continue with clones 2 and 7 in this example because the *Tn7-UAU1* cassette inserted approximately into the middle of the 1800 bp ORF. For a large-scale disruption procedure, we recommend picking clones with middle insertions because this allows for much flanking homology for transformation into *C. albicans*. More homology ensures that the disrupted ORF will more successfully homologously recombine to the native locus of said ORF in *C. albicans*. Other insertion sites may also be of interest for particular ORFs. If so, we recommend sequencing from one transposon junction using Primer S (New England Biolabs), followed by BLASTN analysis against The *Candida* Genome Database at www.candidagenome.org.

15. The doubling time of *C. albicans* is approximately 1.5 h.
16. PEG is cytotoxic, thus incubating for more than 16 h significantly reduces the transformation efficiency.
17. We recommend dividing a YPD plate into quarters, and thickly patching the single colony onto one quarter of the YPD plate. Due to the fact that the gene conversion event transferring the *UAU1* cassette to the other allele occurs very infrequently, a thick patch of cells is necessary.

Acknowledgments

This work was supported by NIH grants R01 AI067703 and T32 DK007786.

References

1. Fonzi, W.A., and Irwin, M.Y. (1993) Isogenic strain construction and gene mapping in Candida albicans. *Genetics* **134**, 717–728.
2. Enloe, B., Diamond, A., and Mitchell, A.P. (2000) A single-transformation gene function test in diploid Candida albicans. *J. Bacteriol.* **182**, 5730–5736.
3. Davis, D.A., Bruno, V.M., Loza, L., Filler, S.G., and Mitchell, A.P. (2002). Candida albicans Mds3p, a conserved regulator of pH responses and virulence identified through insertional mutagenesis. *Genetics* **162**, 1573–1581.
4. Wilson, R.B., Davis, D., and Mitchell, A.P. (1999) Rapid hypothesis testing with Candida albicans through gene disruption with short homology regions. *J. Bacteriol.* **181**, 1868–1874.
5. Davis, D., Wilson, R.B., and Mitchell, A.P. (2000) RIM101-dependent and-independent pathways govern pH responses in Candida albicans. *Mol. Cell. Biol.* **20**, 971–978.

Part VI
Appendix

Chapter 18

Standard Growth Media and Common Techniques for Use with *Candida albicans*

Neeraj Chauhan and Michael D. Kruppa

1. Introduction

There are a variety of standard media and routine procedures commonly used in investigations concerning *Candida albicans* and other fungi. It can, however, be difficult to locate appropriate formulations and protocols, as these are scattered about the literature. Thus, the purpose of this chapter is to provide this information in a convenient manner.

2. Commonly Used Media

1. *YPD (also known as YEPD)*. Glucose of 20 g (2%), 20 g (2%) peptone, 10 g (1%) yeast extract; adjust volume to 1 L with deionized water. For solid medium, add 20 g agar (2%). Sterilize by autoclaving.

2. *M199 (Medium 199) (1)*. M199 of 9.5 g with Earle salts powder (Invitrogen or Mediatech), 18.7 g Trisbase (Trizma, Sigma), add 800 mL of deioinzed water, and adjust pH to 7.5 or 3.5 with HCl as needed. Final volume should be adjusted to 1 L if used as a liquid medium and then filter sterilized. If used as a solid medium, filter sterilize 800 mL of the medium as described, and add 200 mL of autoclaved agar preparation (20 g). Mix before cooled and pour plates.

3. *YNB (yeast nitrogen base)*. Yeast nitrogen base of 6.7 g (0.67%) with ammonium sulfate without dextrose or amino

acids. Glucose of 20 g (2%) is added. Add distilled water to 1 L and autoclave. For solid medium, add 20 g (2%) agar.

4. *YNB and YPD + uridine.* Add 25 μg of uridine per milliliter of media, i.e., 25 μL of 10 mg/mL stock of uridine prepared in water and filter sterilized, store at 4°C.

5. *5-Fluoroorotic acid (5-FOA) (2).* FOA 1 mg per milliliter of media is used for counter selection of the *URA3* gene, in gene disruption experiments. It does not dissolve easily in water but should be added to media even when not completely in solution. Also, be sure that there is uridine present in the medium to allow growth of culture on agar.

6. *SLAD (synthetic low ammonium dextrose) (3).* YNB 1.7 g (0.17%) w/o amino acids and w/o ammonium sulfate, 20 g (2%) glucose, 50 μM ammonium sulfate, 20 g (2%) agar. Add distilled water to 1 L and sterilize by autoclaving.

7. *Serum agar media (4).* Serum 10% from calf, horse, pig, or sheep, 20 g (2%) agar. Add agar to 900 mL water, autoclave, and allow to cool. Subsequently, add 100 mL serum. Pour plates. Alternatively, serum can be added to 900 mL of YPD medium.

8. *Spider medium (5).* Nutrient broth of 10 g, 10 g D-mannitol, 2 g K_2HPO_4, 20 g agar. Add distilled water to 1 L and sterilize by autoclaving.

9. *Lee's Medium (6).* Prepare separately in two flasks. (a) Solution 1: 10 g $(NH_4)_2SO_4$, 0.4 g $MgSO_4 \cdot 7H_2O$, 5 g K_2HPO_4, 10 g NaCl, 20 g agar (if solid medium is desired). Adjust pH to 7.0 with HCL and bring volume to 1 L. Autoclave. (b) Solution 2: 1 g L-alanine, 2.6 g L-leucine, 2 g L-lysine, 0.2 g L-methionine, 0.142 g L-ornithine, 1 g L-phenylananline, 1 g L-threonine, 1 g L-proline. Dissolve in 1 L of water. Autoclave. Let cool and then add 25 mL dextrose (20% solution sterilized by filtration), 4 mL biotin (1 mg/mL sterilized by filtration), 28 mL arginine (5 mg/mL sterilized by filtration). Combine solutions and pour plates. Smaller total volumes can be made by adjusting concentrations of components accordingly.

3. Methods

3.1. Cell Counting and Cell Density Determination

Direct cell counting is often used for determining *C. albicans* yeast cell concentration, i.e., cells/mL in a culture. Procedure is described for determining cell density of an overnight culture, but can be used in any circumstance where a cell count is needed.

1. Inoculate 10 mL of an overnight culture in YPD or other media. Grow overnight at 30°C.
2. Harvest the cells by centrifugation and resuspend in an equal volume of sterile distilled water, or other appropriate solution.
3. Remove a 10-µL aliquot of cells and dilute into 990 µL of water or saline. Vortex, and take 10 µL and place on a hemocytometer. Count cells in six grids and determine the average number of cells/grid. To calculate cells/mL:
   Average cell no. per grid $\times$ 100 (dilution factor) $\times 25 \times 10^4 =$ cells/mL
   This will give the number of cells per milliliter in the original preparation (see **Note 1**).

3.2. DNA Isolation (7)

1. Harvest cells (10 mL overnight culture) by centrifugation at 3000 $\times$ g for 10 min.
2. Wash cells with sterile water to remove residual media.
3. Resuspend cell pellet in 200 µL breaking buffer (2% Triton X-100, 1% SDS, 100 mM NaCl, 10 mM Tris–HCL, pH 8.0, and 1 mM EDTA, pH 8.0). Add 0.25 g glass beads (glass beads should be just below the meniscus) and add 200 µL phenol/chloroform/isoamyl alcohol and vortex at maximum speed for 2–3 min.
4. Add 200 µL TE buffer (100 mM Tris–HCl, pH 8.0 and 10 mM EDTA, pH 8.0) and vortex to mix.
5. Microfuge at full speed for 5 min at room temperature and transfer top aqueous layer to a new tube. Add 1 mL 100% ethanol and mix to precipitate DNA.
6. Microfuge at full speed for 5 min at room temperature. Discard supernatant and resuspend pellet in 0.5 mL TE buffer.
7. Add 30 µL of 1 mg/mL RNase A and incubate at 37°C for 15 min.
8. Add 10 µL of 3 M sodium acetate pH 5.5 and 1 mL of 100% ethanol. Mix to precipitate.
9. Microfuge at full speed for 5 min at room temperature. Remove supernatant and air-dry pellet. Resuspend pellet in 100 µL TE buffer. Quantitate by taking OD at 260 nm.

3.3. RNA Isolation

For the isolation of total RNA from *C. albicans*, several methods are available. However, the hot phenol method (8) is one of the most widely used methods. Throughout the protocol, strict care must be taken to minimize contamination with RNAses.

1. Harvest cells (approximately 10 mL of an overnight culture equivalent to approximately 5–7 OD_{600} units) by centrifugation at 3000 $\times$ g for 10 min.

2. Wash cell pellet twice with DEPC-treated water and store frozen at −20°C until use.
3. Resuspend cell pellet in 400 µL of TES (10 mM Tris–HCl, pH 7.5, 10 mM EDTA, 0.5% SDS), and add 400 µL of acid phenol to the mix.
4. Incubate the tubes at 65°C for 30–60 min with occasional brief vortexing.
5. Place on ice for 5 min; microfuge at 13,000 × g for 5 min.
6. Transfer aqueous phase to a new tube and add 400 µL acid phenol, vortex, and keep in ice for 5 min and microfuge at 13,000 × g for 5 min.
7. Transfer aqueous phase to a new tube and add 400 µL of chloroform. Vortex and centrifuge at 13,000 × g for 5 min.
8. Precipitate supernatant with 40 µL 3 M sodium acetate pH 5.3 and 1 mL of ice-cold 100% ethanol. Wash pellet with ice-cold 70% ethanol.
9. Dissolve RNA in DEPC-treated water and determine its concentration by using a spectrophotometer at A_{260} and A_{280}.

3.4. Suggested Gene Nomenclature

For a general overview and instructions for gene nomenclature that should be followed for *Candida* genes, the reader is referred to *C. albicans* gene nomenclature guide available online at the Candida genome database (CGD-www.candidagenome.org).

4. Notes

1. Use following method for counting cells.

    Count cells within the six squares in a hemocytometer. Count all squares in the counting grid (25 total). For determining yeast cells per milliliter, use following formula:

    Yeast cells/mL = average number of cells in six diagonal squares × dilution factor (100) × 25 (all squares in the grid) × 10^4 (or 10,000 this factor incorporates the volume of fluid in the counting chamber).

References

1. Kimura, L. H. and Pearsall, N. N. (1978) Adherence of *Candida albicans* to human buccal epithelial cells. *Infect. Immun.* **21**, 64–68.
2. Boeke, J. D., Lacroute, F. and Fink, G. R. (1984) A positive selection for mutants lacking orotidine-5'-phosphate decarboxylase activity in yeast: 5-fluoro-orotic acid resistance. *Mol. Gen. Genet.* **197**, 345–346.
3. Gimeno, C. J. and Fink, G. R. (1994) Induction of pseudohyphal growth by overexpression of *PHD1*, a *Saccharomyces cerevisiae* gene related to transcriptional regulators of fungal development. *Mol. Cell. Biol.* **14**, 2100–2112.

4. Mackenzie, D. W. R. (1962) Serum tube identification of *Candida albicans*. *J. Clin. Path.* **15**, 563–565.
5. Liu, H., Köhler, J. and Fink, G. (1994) Suppression of hyphal formation in *Candida albicans* by mutation of a STE12 homolog. *Science* **266**, 1723–1726.
6. Lee, K. L., Buckley, H. R. and Campbell, C. C. (1975). An amino acid liquid synthetic medium for the development of mycelial and yeast forms of *Candida albicans*. *Sabouraudia* **13**, 148–153.
7. Cryer, D. R., Eccleshall, R. and Marmur, J. (1975) Isolation of yeast DNA. *Methods Cell Biol.* **12**, 39–44.
8. Köhrer, K. and Domdey, H. (1991) Preparation of high molecular weight RNA. *Methods Enzymol.* **194**, 398–405.

INDEX

A

Als protein 27, 28, 30, 31, 32
    ab ELISA 28–29, 30, 31
    ag vaccine 30, 31, 32
    r-Als protein 30
    vaccine 31, 32
Amphotericin B 38, 43, 55, 56, 57, 58, 59
Animal infection model, see *Candida albicans*, mouse infection model
Antibodies 9, 10, 14, 15, 27
    detection of 9
    mucosal surfaces 9
    r-Als 28, 30, 31
    reporter 13
    sIgA 9, 10, 11, 14
Antifungal agents 37, 38, 40, 41, 42–43
    biofilm experiments 38, 39, 41, 55, 56
Antigens 3, 4, 9, 13, 15, 27, 28, 29, 30
    concentration 12
    ELISA 12–14
    r-Als 28, 30, 31, 32
Artificial throat 45, 46, 47, 50
    experiments with 47–48, 49
    FISH 47
    microbial species in biofilms 48, 50

B

Biofilm 37, 45, 55, 56, 57
    antifungal susceptibility assay 55, 56, 57, 58
    C. albicans model systems 46–47
    drug resistance 37–38
    formation in microtiter plates 55, 56–61
    identification of species 47
    inoculum 38, 39
    in vitro adhesion 57
    in vitro formation 38, 39
    in vitro viability 58–59
    in vivo formation 45–46
    microtiter plate assay 56–61
    mixed-species 45
    penetration by antifungal agents 38–42, 43
    shunt prostheses 46, 49
    viable cell counts 39, 41
Biotyping 97, 98, 99, 102, 103, 112
    See also Yeast killer system
Blocking 13, 15
    ELISA 13
    reagent 11
Body fluids 9
    types 9

C

Candida 12, 17, 18, 19, 20, 22,
    23, 24, 28, 29, 37, 38, 39, 43, 46, 55, 66, 68, 77,
    78, 80, 81, 90, 97, 115, 126, 149, 154, 158, 159
    antigens 11, 12, 13, 14, 15
    biofilm 46, 55, 57
    blastoconidia prep 22
    cell culture 119
    cells 11, 12, 13, 14, 15, 22, 23, 25, 68, 69
    cell wall 150
    co-culture with phagocytes 22–23
    DNA purification 121–122
    ELISA 12–14
    FITC-labelled 19, 23, 25
    genetic relatedness 115, 118
    in vivo cultures 78
    innate immunity 17
    RNA extraction 79, 81
    sIgA antibodies 9, 10, 11
Candida albicans 9, 31, 32, 33,
    37, 39, 40, 45, 46, 65, 66, 69, 73, 74, 75, 77, 78,
    82, 83, 85, 87, 88, 90, 97, 99, 103, 109, 115, 125,
    131, 132, 133, 139, 141, 142, 143, 147, 149, 155,
    156, 157, 159, 161, 162, 163, 167, 170, 173, 175,
    179, 180, 187, 190, 191, 192, A1, A2, A33
    antibodies 9, 10, 11, 12, 15
    antigens 11, 12, 13, 14, 15
    biofilm 45, 46, 55, 56, 59

(continued)
- biofilm model systems ... 46–47
- cell counting ... A2-A3, A44
- cell density ... A2-A3
- cell growth ... 137, 139
- cell lysis ... 137–138, 139–140
- cells ... 11, 12, 13, 14, 15
- cell wall ... 149, 150
- cell wall protein isolation ... 150, 151–152
- cell wall proteomic analysis ... 150–155
- commonly used media ... A1-A22
- DNA isolation ... A3
- gene disruption ... 175, 176, 177–194
- genetic transformation of ... 169–173
- genomic DNA prep ... 177–178, 180, 181, 193
- homozygous disruption mutants ... 181, 190–192
- inoculation ... 78
- *in vitro* culture harvest ... 81, 139
- *in vitro* tissue model ... 85, 87–91
- *in vivo* gene expression model ... 77, 78–82
  See also mouse infection model
- luciferase reporter system ... 158
- mouse infection model ... 77, 78, 79
- mouse sample harvest ... 78, 80–81
- non-murine virulence models ... 85–87
- r-Als antigen ... 30, 31, 32
- reporter systems ... 157–159
- RNA extraction ... 79, 81, A3-A44
- strain BWP17 ... 178, 186, 191
- strain CAF2-1 ... 87, 88, 150
- strain CAI4 ... 87, 92
- strain DAY185 ... 193
- strain DAY286 ... 193
- strain HB12 ... 56, 59, 61
- strain SC5314 ... 31, 56, 61, 66, 68, 74, 75, 76, 87, 88, 150
- TAP-tag ... 133, 134
- transformation ... 175, 176, 177, 178, 180, 183, 188–190, 193, 194
- typing systems ... 102–103
- virulence ... 65, 66, 67, 164

Candida Genome Database ... 173, 191, A44
Candidiasis ... 18, 20, 27, 65–66
- clinical manifestations ... 27–28, 29, 30
- immunoprotection ... 27
- mouse models ... 28, 29, 30, 65, 66, 67

CLSM
- FISH analysis ... 47, 49, 50
- visualization of Candida biofilms ... 57, 59–61

Confocal laser scanning microscopy, *see* CLSM

D

Dendritic cells ... 3–4
- continuous cell lines ... 7
- fungal infections ... 4
- generation from PBMCs *in vitro* ... 4–8

DNA fingerprinting ... 115, 118, 121, 122
- methods ... 118
- probe ... 120, 122, 126

Drosophila melanogaster
- inoculation ... 91–92
- virulence model ... 85, 87

E

ELISA ... 10, 15, 24, 29, 30, 31
- assay ... 10–11
- r-Als ... 28–30, 31
- sIgA antibody detection ... 12

Enzyme-linked immunosorbent assay, *see* ELISA

F

Filter disk assay ... 37, 38, 39–42
Firefly luciferase system ... 158, 166
- assay ... 160, 163, 165

FISH ... 47
- artificial throat model ... 47
- assay ... 48, 50
- biofilm species identification ... 47
- CLSM ... 47, 49, 50
- probes ... 48

Fluorescent *in situ* hybridization, *see* FISH

G

β-galactosidase reporter system ... 157–158
- assay ... 159–160, 165

Gene disruption, *see* Large-scale gene disruption
Genetic transformation ... 169
- cation-induced ... 169, 170–173
- electroporation ... 170
- spheroplast method ... 169–170

GFP reporter system ... 157, 158–159, 163–164, 165, 166
- promoter analysis ... 59, 61, 165

I

Immunization ... 27, 28
- adjuvants ... 28, 29
- efficacy ... 28–29, 30
- rAls vaccine ... 31, 32
- routes of ... 28, 29–30

Intraspecific differentiation ... 97–98

of *C. albicans* 99, 101
In vitro tissue model 85, 87, 88–90
    transcriptional study.. 88, 90–91

K

Ketoconazole.. 55, 56, 58, 59
Killer toxins... 97, 98, 102, 104,
    106, 107, 109, 110, 112
    killer toxin based system 108–110
    preparation of... 105–106, 112
    See also Yeast killer system
Killer yeasts 97, 98, 99, 101, 102,
    103, 104, 105, 106, 109, 112, 113
    killer yeast based system................................... 107, 109
    See also Yeast killer system
Killing assays... 19, 22, 23, 24
    colony-forming unit method 23–24
    XTT assays, *see* XTT

L

LacZ reporter system 157–158, 166
    assay... 160–162, 165
Large-scale gene disruption.................... 175, 176, 177–194

M

Mass spectrometry analysis.. 153
    MALDI-TOF-MS 154
Microarray.. 77, 79, 82
    data analysis 79, 82
    sample labeling 79, 82
Microtiter plate assay 56–61
MLST 115, 118, 127, 128
    protocol 120–121, 124–126
Monocytes.. 17, 18, 20
    differentiation into dendritic cells 6
    isolation of 4–5
    See also phagocytic cells
Mucosal secretions, 9
    collection of.. 10, 12
Mucosal surfaces 9, 17, 29, 32
Multilocus sequence typing, *see* MLST
Murine candidiasis models 28, 29, 30, 65, 66
    disseminated candidiasis 65, 66, 68,
    69–72, 73, 75
    oral/esophageal candidiasis 65, 66, 68–69, 73, 76

N

Nonanimal virulence models
    ex vivo 85
    See also D. melanogaster
    in vitro... 85, 87–91

P

pGEMT-Easy plasmid........................ 178, 183, 184, 185,
    186, 187, 189, 192, 193
    ligation 178, 183, 193
    pGEMT-Easy-ORF 178, 184, 186, 187
Phagocytic cells 17
    cell lines.. 18, 20, 21
    monocytes 17–18
    See also PMNs
Phagocytosis.. 17, 18
    assay... 19–21, 22–23, 24, 25
PMNs.. 17, 18, 25
    assays 17, 18
    co-culture with yeast.. 22–23
    isolation of 18–19
    killing assay 19, 23–24
    lysis 25
    preparation of... 19–21
Polymorphonuclear neutrophils, *see* PMNs
Promoter analysis.. 157, 158, 165
    C. albicans reporter systems...................................... 157
    See also reporter systems
Proteomic analysis
    C. albicans cell wall... 150–155
Proteomics 149, 150

R

Renilla luciferase system 156
    assay... 160, 162–163
    See also RLUC
Reporter systems 157–159
    gene-specific constructs 159
    uses of... 157–159
RLUC 158, 160, 162

S

Scanning electron microscopy 39, 42, 87, 88
SDS-PAGE.. 142–143,
    146, 152
Secretory immunoglobulin A, *see* sIgA
Shunt prostheses 45, 46, 48, 49, 50
    Candida biofilm formation 46
    types of... 46, 48, 50
SIgA.. 9, 13, 14, 15
    antibodies 9, 10, 11, 14
    immunoglobulins 10
    standard curve 11, 14
Southern blot hybridization................... 115, 118, 121, 126
    protocol 119–120, 122–124

T

Tandem-affinity purification, *see* TAP
TAP 133
TAP-tag 133, 134, 136, 139, 143
    adding to protein 135–137
    analysis of purified proteins 141–143
    purification of tagged proteins 137, 138,
        140–141, 146
    vectors 135–137
Tn7 transposon 175, 178, 180, 181, 184, 185,
        186, 187, 188, 189, 192, 193
    genetic disruption procedure 177–194
    transformation into *C. albicans* 180
    triplication byproduct 177, 192
Transposon mutagenesis 178, 181, 184
Two-dimension PAGE 150–151, 152

U

UAU1 cassette 173, 174, 175, 178, 179,
        182, 183, 184, 185, 186, 187, 190, 191, 192
    See also Tn7 transposon
URA3 .... 74, 89, 92, 93, 135, 137,
        143, 158, 159, 163, 164, 165, 176, 177, 192
    ura+ 176, 177, 190, 193
    ura- 89, 172, 177, 192, 193

V

Vaccines 27, 29
    assessment of efficacy 29

Virulence 85–87
    of *C. albicans* 65, 66, 67, 164
    genetic basis of 175
    mouse models of 77, 78–82
    nonanimal models 85, 87–93
    studies of 73, 87–93
    virulence factors 17, 124
    virulence phenotype 135

W

Western blot 152–153

X

XTT 19, 57, 61
    killing assay 19, 24, 25

Y

Yeast killer phenomenon 98–99
Yeast killer system 97, 99–103, 108, 112, 113
    killer yeast 99, 103, 104,
        105, 106, 107, 112, 113
    killing assay 104
    modified killer system 106–107, 110–112
    mold strain differentiation 106
    yeast killer toxins 102, 105, 106, 107,
        108, 109, 110, 112, 113
    yeast strain differentiation 105–106
    yeasts and molds used 103–105